QUANTUM CRYPTOGRA
SECRET-KEY DISTILL

Quantum cryptography (or quantum key distribution) is a state-of-the-art technique that exploits the properties of quantum mechanics to guarantee the secure exchange of secret keys. This self-contained text introduces the principles and techniques of quantum cryptography, setting it in the wider context of cryptography and security, with specific focus on secret-key distillation.

The book starts with an overview chapter, progressing to classical cryptography, information theory (classical and quantum), and applications of quantum cryptography. The discussion moves to secret-key distillation, then privacy amplification and reconciliation techniques, concluding with the security principles of quantum cryptography. The author explains the physical implementation and security of these systems, and enables engineers to gauge the suitability of quantum cryptography for securing transmission in their particular application.

With its blend of fundamental theory, implementation techniques, and details of recent protocols, this book will be of interest to graduate students, researchers, and practitioners, in electrical engineering, physics, and computer science.

GILLES VAN ASSCHE received his Ph.D. in Applied Sciences from the Center for Quantum Information and Communication at the University of Brussels in 2005. He currently works in the Smartcard ICs Division at STMicroelectronics in Belgium. His research interests include quantum cryptography, classical cryptography, and information theory.

QUANTUM CRYPTOGRAPHY AND SECRET-KEY DISTILLATION

GILLES VAN ASSCHE

CAMBRIDGE
UNIVERSITY PRESS

CAMBRIDGE UNIVERSITY PRESS
Cambridge, New York, Melbourne, Madrid, Cape Town,
Singapore, São Paulo, Delhi, Mexico City

Cambridge University Press
The Edinburgh Building, Cambridge CB2 8RU, UK

Published in the United States of America by Cambridge University Press, New York

www.cambridge.org
Information on this title: www.cambridge.org/9781107410633

© Cambridge University Press 2006

This publication is in copyright. Subject to statutory exception
and to the provisions of relevant collective licensing agreements,
no reproduction of any part may take place without the written
permission of Cambridge University Press.

First published 2006
First paperback edition 2012

A catalogue record for this publication is available from the British Library

ISBN 978-0-521-86485-5 Hardback
ISBN 978-1-107-41063-3 Paperback

Cambridge University Press has no responsibility for the persistence or
accuracy of URLs for external or third-party internet websites referred to in
this publication, and does not guarantee that any content on such websites is,
or will remain, accurate or appropriate.

Contents

Foreword N. J. Cerf and S. W. McLaughlin		*page* ix
Preface		xi
Acknowledgments		xiii
1	**Introduction**	1
	1.1 A first tour of quantum key distribution	4
	1.2 Notation and conventions	12
2	**Classical cryptography**	15
	2.1 Confidentiality and secret-key ciphers	15
	2.2 Secret-key authentication	26
	2.3 Public-key cryptography	29
	2.4 Conclusion	33
3	**Information theory**	35
	3.1 Source coding	35
	3.2 Joint and conditional entropies	40
	3.3 Channel coding	41
	3.4 Rényi entropies	43
	3.5 Continuous variables	45
	3.6 Perfect secrecy revisited	46
	3.7 Conclusion	48
4	**Quantum information theory**	49
	4.1 Fundamental definitions in quantum mechanics	49
	4.2 Qubits and qubit pairs	52
	4.3 Density matrices and quantum systems	54
	4.4 Entropies and coding	55
	4.5 Particularity of quantum information	56
	4.6 Quantum optics	58

	4.7	Conclusion	60
5	**Cryptosystems based on quantum key distribution**		**63**
	5.1	A key distribution scheme	63
	5.2	A secret-key encryption scheme	70
	5.3	Combining quantum and classical cryptography	73
	5.4	Implementation of a QKD-based cryptosystem	77
	5.5	Conclusion	84
6	**General results on secret-key distillation**		**85**
	6.1	A two-step approach	85
	6.2	Characteristics of distillation techniques	87
	6.3	Authenticated one-shot secret-key distillation	88
	6.4	Authenticated repetitive secret-key distillation	92
	6.5	Unauthenticated secret-key distillation	96
	6.6	Secret-key distillation with continuous variables	98
	6.7	Conclusion	100
7	**Privacy amplification using hash functions**		**101**
	7.1	Requirements	101
	7.2	Universal families suitable for privacy amplification	104
	7.3	Implementation aspects of hash functions	107
	7.4	Conclusion	112
8	**Reconciliation**		**113**
	8.1	Problem description	113
	8.2	Source coding with side information	116
	8.3	Binary interactive error correction protocols	124
	8.4	Turbo codes	129
	8.5	Low-density parity-check codes	137
	8.6	Conclusion	140
9	**Non-binary reconciliation**		**141**
	9.1	Sliced error correction	141
	9.2	Multistage soft decoding	148
	9.3	Reconciliation of Gaussian key elements	149
	9.4	Conclusion	158
10	**The BB84 protocol**		**159**
	10.1	Description	159
	10.2	Implementation of BB84	160
	10.3	Eavesdropping and secret key rate	170
	10.4	Conclusion	181

11	**Protocols with continuous variables**	183
	11.1 From discrete to continuous variables	183
	11.2 A protocol with squeezed states	184
	11.3 A protocol with coherent states: the GG02 protocol	189
	11.4 Implementation of GG02	194
	11.5 GG02 and secret-key distillation	198
	11.6 Conclusion	203
12	**Security analysis of quantum key distribution**	205
	12.1 Eavesdropping strategies and secret-key distillation	205
	12.2 Distillation derived from entanglement purification	207
	12.3 Application to the GG02 protocol	221
	12.4 Conclusion	244
Appendix: symbols and abbreviations		245
Bibliography		249
Index		259

Foreword

The distribution of secret keys through quantum means has certainly become the most mature application of quantum information science. Much has been written on quantum cryptography today, two decades after its inception by Gilles Brassard and Charles Bennett, and even longer after the pioneering work of Stephen Wiesner on non-counterfeitable quantum money which is often considered as the key to quantum cryptography. Quantum key distribution has gone from a bench-top experiment to a practical reality with products beginning to appear. As such, while there remain scientific challenges, the shift from basic science to engineering is underway. The wider interest by both the scientific and engineering community has raised the need for a fresh new perspective that addresses both.

Gilles Van Assche has taken up the challenge of approaching this exciting field from a non-traditional perspective, where classical cryptography and quantum mechanics are very closely intertwined. Most available papers on quantum cryptography suffer from being focused on one of these aspects alone, being written either by physicists or computer scientists. In contrast, probably as a consequence of his twofold background in engineering and physics, Gilles Van Assche has succeeded in writing a comprehensive monograph on this topic, which follows a very original view. It also reflects the types of challenge in this field – moving from basic science to engineering. His emphasis is on the classical procedures of authentication, reconciliation and privacy amplification as much as on the quantum mechanical basic concepts. Another noticeable feature of this book is that it provides detailed material on the very recent quantum cryptographic protocols using continuous variables, to which the author has personally contributed. This manuscript, which was originally written as a dissertation for the author's Ph.D. thesis, is excellent and, hence, was very appropriate to be turned into the present book.

After an introduction to the basic notions of classical cryptography, in particular secret-key ciphers and authentication together with the concept of information-theoretic security, the tools of quantum information theory that are needed in the present context are outlined in the first chapters. The core of the book starts with Chapter 5, which makes a thorough description of quantum cryptosystems, from the theoretical concepts to the optical implementation. Chapter 6 considers the classical problem of distilling a secret key from the quantum data, a topic which is rarely treated to this depth in the current literature. The implementation of privacy amplification and reconciliation is illustrated more specifically in Chapters 7 and 8, while the case of continuous-variable reconciliation, which is the central contribution of Gilles Van Assche's thesis, is treated in Chapter 9. Then, the last chapters of the book study discrete-variable and continuous-variable quantum cryptographic protocols and analyze their security.

Gilles Van Assche has produced a remarkably self-contained book, which is accessible to newcomers to the field with a basic background in physical and computer sciences, as well as electrical engineering. Being fully up-to-date, this book will, at the same time, prove very useful to the scientists already involved in quantum cryptography research. With its science and engineering perspective, this book will undoubtedly become a reference in this field.

NICOLAS J. CERF
Professor
Université Libre de Bruxelles

STEVEN W. MCLAUGHLIN
Ken Byers Distinguished Professor
Georgia Institute of Technology

Preface

This book aims at giving an introduction to the principles and techniques of quantum cryptography, including secret-key distillation, as well as some more advanced topics. As quantum cryptography is now becoming a practical reality with products available commercially, it is important to focus not only on the theory of quantum cryptography but also on practical issues. For instance, what kind of security does quantum cryptography offer? How can the raw key produced by quantum cryptography be efficiently processed to obtain a usable secret key? What can safely be done with this key? Many challenges remain before these questions can be answered in their full generality. Yet quantum cryptography is mature enough to make these questions relevant and worth discussing.

The content of this book is based on my Ph.D. thesis [174], which initially focused on continuous-variable quantum cryptography protocols. When I decided to write this book, it was essential to include discrete-variable protocols so as to make its coverage more balanced. In all cases, the continuous and discrete-variable protocols share many aspects in common, which makes it interesting to discuss about them both in the same manuscript.

Quantum cryptography is a multi-disciplinary subject and, in this respect, it may interest readers with different backgrounds. Cryptography, quantum physics and information theory are all necessary ingredients to make quantum cryptography work. The introductory material in each of these fields should make the book self-contained. If necessary, references are given for further readings.

Structure of this book

The structure of this book is depicted in Fig. 0.1. Chapter 1 offers an overview of quantum cryptography and secret-key distillation. Chapters 2,

3 and 4 give some aspects of classical cryptography, classical information theory and quantum information theory, respectively.

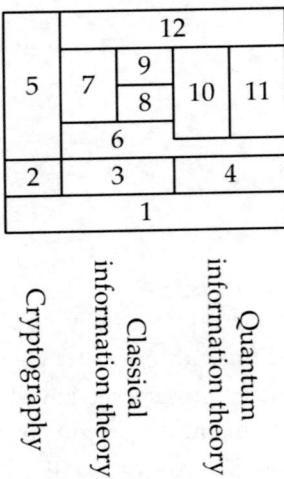

Fig. 0.1. Dependencies between chapters: a chapter depends on the chapter or chapters beneath. Block sizes are arbitrary.

Chapters 5–11 follow a top-down approach. First, Chapter 5 discusses quantum cryptography from an application point of view and macroscopically describes what services it provides and what are its prerequisites. Confidentiality requires a secret key, and Chapter 6 shows how to obtain one with secret-key distillation techniques. Secret-key distillation is further detailed in Chapters 7–9. Chapter 7 explains how to make the key secret using privacy amplification. This in turn requires the key to be error-free, and in this respect, the reconciliation techniques are detailed in Chapters 8 and 9. Then, the quantum sources of key elements to distill are described in Chapter 10 for discrete variables and in Chapter 11 for continuous variables.

Finally, Chapter 12 analyzes the security principles of quantum cryptography and revisits secret-key distillation from a quantum-physical perspective.

Error reporting

If you find any error in this book, please do not hesitate to report it. You can find the contact information and an errata list on the web page: http://gva.noekeon.org/QCandSKD/.

Acknowledgments

This book would not have been written without the support and help of many people. In particular, I would like to thank:

- Nicolas Cerf, my thesis supervisor, for his advice and support during the thesis;
- Steven McLaughlin, for his strong encouragements to take on this project and for his suggestions;
- the remaining members of the committee for their enthusiastic feedback: Daniel Baye, Michel Collard, Philippe Grangier, Olivier Markowitch, Serge Massar, and Louis Salvail;
- all the other researchers, with whom I worked or co-signed articles, or who reviewed parts of the text during the writing of this book: Matthieu Bloch, Jean Cardinal, Joan Daemen, Samuel Fiorini, Frédéric Grosshans, Sofyan Iblisdir, Marc Lévy, Patrick Navez, Kim-Chi Nguyen, Michaël Peeters, Rosa Tualle-Brouri, and Jérôme Wenger;
- Serge Van Criekingen for his thorough proof reading;
- my colleagues at the Center for Quantum Information and Communication for helpful discussion throughout the thesis;
- my colleagues at STMicroelectronics for their encouragements;
- my family and friends for their moral support;
- and last but not least, Céline for her encouragements and patience during the numerous hours I was busy working on this project.

1
Introduction

In the history of cryptography, quantum cryptography is a new and important chapter. It is a recent technique that can be used to ensure the confidentiality of information transmitted between two parties, usually called Alice and Bob, by exploiting the counterintuitive behavior of elementary particles such as photons.

The physics of elementary particles is governed by the laws of quantum mechanics, which were discovered in the early twentieth century by talented physicists. Quantum mechanics fundamentally change the way we must see our world. At atomic scales, elementary particles do not have a precise location or speed, as we would intuitively expect. An observer who would want to get information on the particle's location would destroy information on its speed – and vice versa – as captured by the famous *Heisenberg uncertainty principle*. This is not a limitation due to the observer's technology but rather a fundamental limitation that no one can ever overcome.

The uncertainty principle has long been considered as an inconvenient limitation, until recently, when positive applications were found.

In the meantime, the mid-twentieth century was marked by the creation of a new discipline called *information theory*. Information theory is aimed at defining the concept of information and mathematically describing tasks such as communication, coding and encryption. Pioneered by famous scientists like Turing and von Neumann and formally laid down by Shannon, it answers two fundamental questions: what is the fundamental limit of data compression, and what is the highest possible transmission rate over a communication channel?

Shannon was also interested in cryptography and in the way we can transmit confidential information. He proved that a perfectly secure cipher would need a secret key that is as long as the message to encrypt. But he does not say how to obtain such a long secret key. This is rather limiting because the secret key needs to be transmitted confidentially, e.g., using a diplomatic suitcase. If we had a way, say a private line, to transmit it securely, we could directly use this private line to transmit our confidential information.

Since the seventies and up to today, cryptographers have found several clever ways to send confidential information using encryption. In particular, classical ciphers encrypt messages using a small secret key, much smaller than the message size. This makes confidentiality achievable in practice. Yet, we know from Shannon's theory that the security of such schemes cannot be perfect.

Leaving aside the problem of sending confidential information, let us come back to information theory. Shannon defined information as a mathematical concept. Nevertheless, a piece of information must somehow be stored or written on a medium and, hence, must follow the laws of physics. Landauer was one of the first to realize the consequences of the fact that any piece of information ultimately exists because of its physical support. Shannon's theory essentially assumes a classical physical support. When the medium is of atomic scale, the carried information behaves quite differently, and all the features specific to quantum mechanics must be translated into an information-theoretic language, giving rise to *quantum information theory*.

The first application of quantum information theory was found by Wiesner in the late sixties [186]. He proposed using the spin of particles to make unforgeable bank notes. Roughly speaking, the spin of a particle obeys the uncertainty principle: an observer cannot get all the information about the spin of a single particle; he would irreversibly destroy some part of the information when acquiring another part. By encoding identification information on bank notes in a clever way using elementary particles, a bank can verify their authenticity by later checking the consistency of this identification information. At the atomic scale, the forger cannot perfectly copy quantum information stored in the elementary particles; instead, he will unavoidably make mistakes. Simply stated, copying the bank note identification information is subject to the uncertainty principle, and thus a forgery will be distinguishable from a legitimate bank note.

Other applications of quantum information theory were found. For instance, a *quantum computer*, that is, a computer that uses quantum principles instead of the usual classical principles, can solve some problems much faster than the traditional computer. In a classical computer, every computation is a combination of zeroes and ones (i.e., bits). At a given time, a bit can either be zero or one. In contrast, a *qubit*, the quantum equivalent of a bit, can be a zero and a one at the same time. In a sense, processing qubits is like processing several combinations of zeroes and ones simultaneously, and the increased speed of quantum computing comes from exploiting this parallelism. Unfortunately, the current technologies are still far away from making this possible in practice.

Following the tracks of Weisner's idea, Bennett and Brassard proposed in 1984 a protocol to distribute secret keys using the principles of quantum mechanics called *quantum cryptography* or more precisely *quantum key distribution* [10]. By again exploiting the counterintuitive properties of quantum mechanics, they developed a way to exchange a secret key whose secrecy is guaranteed by the laws of physics. Following the uncertainty principle, an eavesdropper cannot know everything about a photon that carries a key bit and will destroy a part of the information. Hence, eavesdropping causes errors on the transmission line, which can be detected by Alice and Bob.

Quantum key distribution is not only based on the principles of quantum physics, it also relies on classical information theory. The distributed key must be both common and secret. First, the transmission errors must be corrected, whether they are caused by eavesdropping or by imperfections in the setup. Second, a potential eavesdropper must know nothing about the key. To achieve these two goals, techniques from classical information theory, collectively denoted as *secret-key distillation*, must be used.

Unlike the quantum computer, quantum key distribution is achievable using current technologies, such as commercially available lasers and fiber optics. Furthermore, Shannon's condition on the secret key length no longer poses any problem, as one can use quantum key distribution to obtain a long secret key and then use it classically to encrypt a message of the same length. The uncertainty principle finds a positive application by removing the difficulty of confidentially transmitting long keys.

State-of-the-art ciphers, if correctly used, are unbreakable according to today's knowledge. Unfortunately, their small key size does not offer any long-term guarantee. No one knows what the future will bring, so if clever advances in computer science or mathematics once jeopardize today's ciphers' security, quantum key distribution may offer a beautiful alternative solution. Remarkably, the security of quantum key distribution is guaranteed by the laws of quantum mechanics.

Furthermore, quantum key distribution guarantees long-term secrecy of confidential data transmission. Long-term secrets encrypted today using classical ciphers could very well become illegitimately decryptable in the next decades. There is nothing that prevents an eavesdropper from intercepting an encrypted classical transmission and keeping it until technology makes it feasible to break the encryption. On the other hand, the key obtained using quantum key distribution cannot be copied. Attacking the key means attacking the quantum transmission today, which can only be done using today's technology.

For some authors, quantum cryptography and quantum key distribution are synonymous. For others, however, quantum cryptography also includes other applications of quantum mechanics related to cryptography, such as quantum secret sharing. A large portion of these other applications requires a quantum computer, and so cannot be used in practice. On the other hand, the notion of key is so central to cryptography that quantum key distribution plays a privileged role. Owing to this last comment, we will follow the first convention and restrict ourselves to quantum key distribution in the scope of this book.

1.1 A first tour of quantum key distribution

As already mentioned, quantum key distribution (QKD) is a technique that allows two parties, conventionally called Alice and Bob, to share a common secret key for cryptographic purposes. In this section, I wish to give a general idea of what QKD is and the techniques it involves. The concepts will be covered in more details in the subsequent chapters.

To ensure the confidentiality of communications, Alice and Bob agree on a common, yet secret, piece of information called a key. Encryption is performed by combining the message with the key in such a way that the result is incomprehensible by an observer who does not know the key. The recipient of the message uses his copy of the key to decrypt the message.

Let us insist that it is not the purpose of QKD to encrypt data. Instead, the goal of QKD is to guarantee the secrecy of a distributed key. In turn, the legitimate parties may use this key for encryption. The confidentiality of the transmitted data is then ensured by a chain with two links: the quantum-distributed key and the encryption algorithm. If one of these two links is broken, the whole chain is compromised; hence we have to look at the strengths of both links.

First, how is the confidentiality of the key ensured? The laws of quantum mechanics have strange properties, with the nice consequence of making the eavesdropping detectable. If an eavesdropper, conventionally called Eve, tries to determine the key, she will be detected. The legitimate parties will then discard the key, while no confidential information has been transmitted yet. If, on the other hand, no tapping is detected, the secrecy of the distributed key is guaranteed.

As the second link of the chain, the encryption algorithm must also have strong properties. As explained above, the confidentiality of data is absolutely guaranteed if the encryption key is as long as the message to transmit and is not reused for subsequent messages. This is where quantum key distribution is particularly useful, as it can distribute long keys as often as needed by Alice and Bob.

Let us detail further how QKD works. Quantum key distribution requires a transmission channel on which quantum carriers are transmitted from Alice to Bob. In theory, any particle obeying the laws of quantum mechanics can be used. In practice, however, the quantum carriers are usually photons, the elementary particle of light, while the channel may be an optical fiber (e.g., for telecommunication networks) or the open air (e.g., for satellite communications).

In the quantum carriers, Alice encodes random pieces of information that will make up the key. These pieces of information may be, for instance, random bits or Gaussian-distributed random numbers, but for simplicity of the current discussion, let us restrict ourselves to the case of Alice encoding only zeroes and ones. Note that what Alice sends to Bob does not have to – and may not – be meaningful. The whole point is that an eavesdropper cannot predict any of the transmitted bits. In particular, she may not use fixed patterns or pseudo-randomly generated bits, but instead is required to use "truly random" bits – the meaning of "truly random" in this scope will be discussed in Chapter 5.

During the tranmission between Alice and Bob, Eve might listen to the quantum channel and therefore spy on potential secret key bits. This does not pose a fundamental problem to the legitimate parties, as the eavesdropping is detectable by way of transmission errors. Furthermore, the secret-key distillation techniques allow Alice and Bob to recover from such errors and create a secret key out of the bits that are unknown to Eve.

After the transmission, Alice and Bob can compare a fraction of the exchanged key to see if there are any transmission errors caused by eavesdropping. For this process, QKD requires the use of a public classical authenticated channel, as depicted in Fig. 1.1. This classical channel has two important characteristics, namely, publicness and authentication. It is not required to be public, but if Alice and Bob had access to a private channel, they would not need to encrypt messages; hence the channel is assumed to be public. As an important consequence, any message exchanged by Alice and Bob on this channel may be known to Eve. The authentication feature is necessary so that Alice and Bob can make sure that they are talking to each other. We may think that Alice and Bob know each other and will not get fooled if Eve pretends to be either of them – we will come back on this aspect in Section 5.1.1.

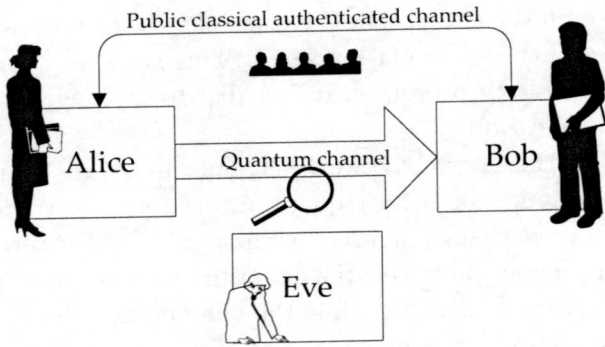

Fig. 1.1. Quantum key distribution comprises a quantum channel and a public classical authenticated channel. As a universal convention in quantum cryptography, Alice sends quantum states to Bob through a quantum channel. Eve is suspected of eavesdropping on the line.

I now propose to overview the first QKD protocol, created by Bennett and Brassard in 1984, called BB84 for short [10]. More than twenty years later, BB84 can still be considered as a model for many other protocols and allows me to introduce the main concepts of QKD.

1.1.1 Encoding random bits using qubits

Any message can, at some point, be converted into zeroes and ones. In classical information theory, the unit of information is therefore the bit, that is, the set $\{0, 1\}$. The quantum carriers of BB84, however, cannot be described in classical terms, so we have to adapt our language to this new setting.

There is a correspondence between the quantum state of some physical system and the information it carries. Quantum states are usually written using Dirac's notation, that is, with a symbol enclosed between a vertical bar and an angle bracket, as in $|\psi\rangle$, $|1\rangle$ or $|x\rangle$; quantum pieces of information follow the same notation.

In quantum information theory, the unit of information is the *qubit*, the quantum equivalent of a bit. Examples of physical systems corresponding to a qubit are the spin of an electron or the polarization of a photon. More precisely, a qubit is described by two complex numbers and belongs to the set

$$\{\alpha|0\rangle + \beta|1\rangle \ : \ |\alpha|^2 + |\beta|^2 = 1, \ \alpha, \beta \in \mathbf{C}\},$$

with $|0\rangle$ and $|1\rangle$ two reference qubits, corresponding to two orthogonal states

1.1 A first tour of quantum key distribution

in a quantum system. The qubits $|0\rangle$ ($\alpha = 1, \beta = 0$) and $|1\rangle$ ($\alpha = 0, \beta = 1$) may be thought of as the quantum equivalent of the bits 0 and 1, respectively. For other values of α and β, we say that the qubit contains a *superposition* of $|0\rangle$ and $|1\rangle$. For instance, the qubits $2^{-1/2}|0\rangle + 2^{-1/2}|1\rangle$ and $\sin \pi/6|0\rangle + i \cos \pi/6|1\rangle$ are both superpositions of $|0\rangle$ and $|1\rangle$, albeit different ones.

In BB84, Alice encodes random (classical) bits, called *key elements*, using a set of four different qubits. The bit 0 can be encoded with either $|0\rangle$ or $|+\rangle = 2^{-1/2}|0\rangle + 2^{-1/2}|1\rangle$. The bit 1 can be encoded with either $|1\rangle$ or $|-\rangle = 2^{-1/2}|0\rangle - 2^{-1/2}|1\rangle$ – note the difference in sign. In both cases, Alice chooses either encoding rule at random equally likely. Then, she sends a photon carrying the chosen qubit to Bob.

When the photon arrives at Bob's station, he would like to decode what Alice sent. For this, he needs to perform a *measurement*. However, the laws of quantum mechanics prohibit Bob from determining the qubit completely. In particular, it is impossible to determine accurately the coefficients α and β of the received qubit $\alpha|0\rangle + \beta|1\rangle$. Instead, Bob must choose a pair of *orthogonal* qubits and perform a measurement that distinguishes only among them. We say that two qubits, $|\phi\rangle = \alpha|0\rangle + \beta|1\rangle$ and $|\psi\rangle = \alpha'|0\rangle + \beta'|1\rangle$, are orthogonal iff $\alpha\alpha'^* + \beta\beta'^* = 0$.

Let us take for instance the qubits $|0\rangle$ and $|1\rangle$, which are orthogonal. So, Bob can make a measurement that distinguishes whether Alice sends $|0\rangle$ or $|1\rangle$. But what happens if she sends $|+\rangle$ or $|-\rangle$? Actually, Bob will obtain a result at random! More generally, if Bob receives $|\phi\rangle = \alpha|0\rangle + \beta|1\rangle$ he will measure $|0\rangle$ with probability $|\alpha|^2$ and $|1\rangle$ with probability $|\beta|^2$ – remember that $|\alpha|^2 + |\beta|^2 = 1$. In the particular case of $|+\rangle$ and $|-\rangle$, Bob will get either $|0\rangle$ or $|1\rangle$, each with probability 1/2. Consequently, Bob is not able to distinguish between $|+\rangle$ and $|-\rangle$ in this case and gets a bit value uncorrelated from what Alice sent.

So, what is so special about the qubits $|0\rangle$ and $|1\rangle$? Nothing! Bob can as well try to distinguish any pair of orthogonal states, for instance $|+\rangle$ and $|-\rangle$. Note that $|0\rangle$ and $|1\rangle$ can be equivalently written as $|0\rangle = 2^{-1/2}|+\rangle + 2^{-1/2}|-\rangle$ and $|1\rangle = 2^{-1/2}|+\rangle - 2^{-1/2}|-\rangle$. Hence, in this case, Bob will perfectly decode Alice's key element when she sends $|+\rangle$ and $|-\rangle$, but he will not be able to distinguish $|0\rangle$ and $|1\rangle$. An example of transmission is depicted in Fig. 1.2.

In the BB84 protocol, Bob randomly chooses to do either measurement. About half of the time, he chooses to distinguish $|0\rangle$ and $|1\rangle$; the rest of the time, he distinguishes $|+\rangle$ and $|-\rangle$. At this point, Alice does not reveal which encoding rule she used. Therefore, Bob measures correctly only half of the bits Alice sent him, not knowing which ones are wrong. After sending a long stream of key elements, however, Alice tells Bob which encoding rule

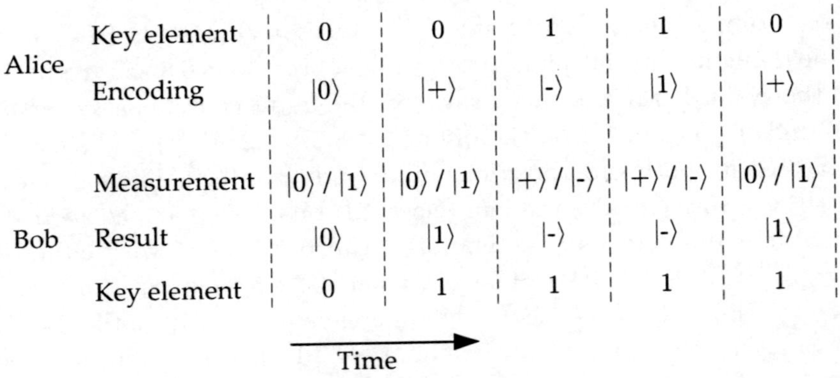

	Key element	0	0	1	1	0
Alice	Encoding	$|0\rangle$	$|+\rangle$	$|-\rangle$	$|1\rangle$	$|+\rangle$
	Measurement	$|0\rangle/|1\rangle$	$|0\rangle/|1\rangle$	$|+\rangle/|-\rangle$	$|+\rangle/|-\rangle$	$|0\rangle/|1\rangle$
Bob	Result	$|0\rangle$	$|1\rangle$	$|-\rangle$	$|-\rangle$	$|1\rangle$
	Key element	0	1	1	1	1

Time →

Fig. 1.2. Example of transmission using BB84. The first two rows show what Alice sends. The bottom rows show the measurement chosen by Bob and a possible result of this measurement.

she chose for each key element, and Bob is then able to discard all the wrong measurements; this part of the protocol is called the *sifting*, which is illustrated in Fig. 1.3.

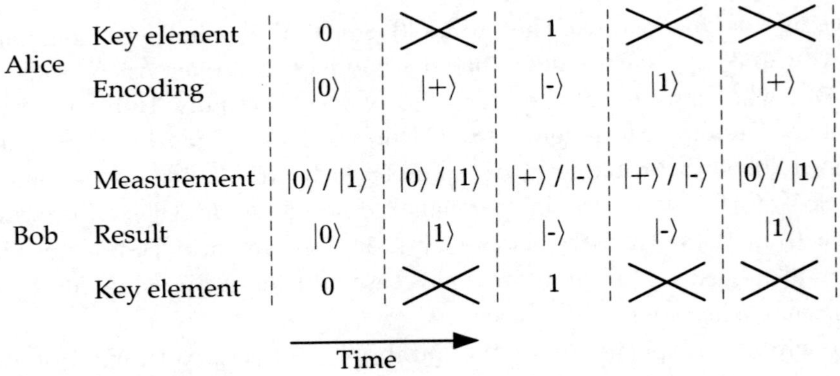

	Key element	0	✗	1	✗	✗
Alice	Encoding	$|0\rangle$	$|+\rangle$	$|-\rangle$	$|1\rangle$	$|+\rangle$
	Measurement	$|0\rangle/|1\rangle$	$|0\rangle/|1\rangle$	$|+\rangle/|-\rangle$	$|+\rangle/|-\rangle$	$|0\rangle/|1\rangle$
Bob	Result	$|0\rangle$	$|1\rangle$	$|-\rangle$	$|-\rangle$	$|1\rangle$
	Key element	0	✗	1	✗	✗

Time →

Fig. 1.3. Sifting of the transmission of Fig. 1.2. The key elements for which Bob's measurement does not match Alice's encoding rule are discarded.

To summarize so far, I have described a way for Alice to send random bits to Bob. Alice chooses among four different qubits for the encoding (two possible qubits per bit value), while Bob chooses between two possible measurement procedures for the decoding. Bob is not always able to determine what Alice sent, but after sifting, Alice and Bob keep a subset of bits for which the transmission was successful. This transmission scheme allows Alice and Bob to detect eavesdropping, and this aspect is described next.

1.1.2 Detecting eavesdropping

The key feature for detecting eavesdropping is that the information is encoded in non-orthogonal qubits. Eve can, of course, intercept the quantum carriers and try to measure them. However, like Bob, she does not know in advance which set of carriers Alice chose for each key element. Like Bob, she may unsuccessfully distinguish between $|0\rangle$ and $|1\rangle$ when Alice encodes a bit as $|+\rangle$ or $|-\rangle$, or vice versa.

In quantum mechanics, measurement is destructive. Once measured, the particle takes the result of the measurement as a state. More precisely, assume that an observer measures a qubit $|\phi\rangle$ so as to distinguish between $|0\rangle$ and $|1\rangle$. After the measurement, the qubit will become either $|\phi\rangle \to |\phi'\rangle = |0\rangle$ or $|\phi\rangle \to |\phi'\rangle = |1\rangle$, depending on the measurement result, *no matter what $|\phi\rangle$ was!* In general, the qubit after measurement $|\phi'\rangle$ is not equal to the qubit before measurement $|\phi\rangle$, except if the qubit is one of those that the observer wants to distinguish (i.e., $|0\rangle$ or $|1\rangle$ in this example).

Every time Eve intercepts a photon, measures it and sends it to Bob, she has a probability 1/4 of introducing an error between Alice's and Bob's bits. Let us break this down. Eve has a probability 1/2 of measuring in the right set. When she does, she does not disturb the state and goes unnoticed. But she is not always lucky. When she measures in the wrong set, however, she sends the wrong state to Bob (e.g., $|+\rangle$ or $|-\rangle$ instead of $|0\rangle$ or $|1\rangle$). This situation is depicted in Fig. 1.4. With the wrong state, Bob will basically measure a random bit, which has a probability 1/2 of matching Alice's bit and a probability 1/2 of being wrong.

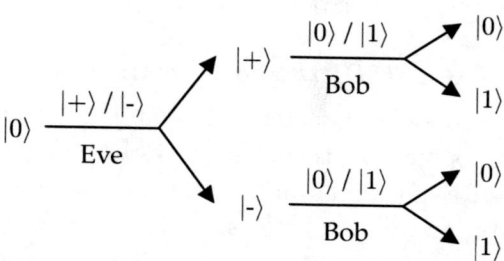

Fig. 1.4. Possible events when Eve uses the wrong measurement for eavesdropping.

So, when Eve tries to eavesdrop, she will get irrelevant results about half of the time and disturb the state. She might decide not to send Bob the states for which she gets irrelevant results, but it is impossible for her to make such a distinction, as she does not know in advance which encoding is used. Discarding a key element is useless for Eve since this sample will not be used by Alice and Bob to make the key. However, if she does retransmit the state (even though it is wrong half of the time), Alice and Bob will detect her presence by an unusually high number of errors between their key elements.

Both Bob and Eve have the same difficulties in determining what Alice sent, since they do not know which encoding is used. But the situation is not symmetric in Bob and Eve: all the communications required to do the sifting are made over the classical authenticated channel. This allows Alice to make sure she is talking to Bob and not to Eve. So, the legitimate parties can guarantee that the sifting process is not influenced by Eve. Owing to this, Alice and Bob can select only the key elements which are correctly measured.

To detect the presence of an eavesdropper, Alice and Bob must be able to detect transmission errors. For this, an option is to disclose a part of the sifted key. A given protocol might specify that after a transmission of $l + n$ key elements (e.g., $l + n = 100\,000$), numbered from 0 to $l + n - 1$, Alice randomly chooses n indexes (e.g., $n = 1000$) and communicates them to Bob. Alice and Bob then reveal the corresponding n key elements to one another so as to count the number of errors. Any error means there was some eavesdropping. The absence of error gives some statistical confidence on the fact that there was no eavesdropping – Eve might just have been lucky, guessing right the encoding sets or making errors only on the other l key elements. Of course, only the remaining l key elements will then be used to produce a secret key.

1.1.3 Distilling a secret key

In the case where errors are detected, Alice and Bob may decide to abort the protocol, as errors may be caused by eavesdropping. At least, this prevents the creation of a key that can be known to the adversary. This kind of decision, however, may be a little stringent. In practice, the physical implementation is not perfect and errors may occur for many reasons other than eavesdropping, such as noise or losses in the quantum channel, imperfect generation of quantum states or imperfect detectors. Also, Eve may just eavesdrop a small fraction of the sifted key, making the remaining key elements available for creating a secret key. There should thus be a way to make a QKD protocol more robust against noise.

Alice and Bob count the number of errors in the disclosed key elements and divide this number by n to obtain an estimate of the expected fraction e

of transmission errors in the whole set of key elements; e is called the *bit error rate*. They can then deduce the amount of information Eve knows about the key elements. For instance, they can statistically estimate that Eve knows no more than, say, I_E bits on the l key elements. This is the *estimation* part of the protocol. The formula giving the quantity I_E is not described here; it results from an analysis of what an eavesdropper may do given the laws of quantum mechanics. Also, the quantity I_E does not precisely tell Alice and Bob what Eve knows about the key elements. She may know the exact value of I_E key elements or merely the result of some arbitrary function of the l key elements, which gives her I_E bits of information in the Shannon sense.

At this point, Alice and Bob know that the l undisclosed key elements have some error rate e and that a potential eavesdropper acquired up to I_E bits of information on them. Using the public classical authenticated channel, Alice and Bob can still try to make a fully secret key; this part is called *secret-key distillation*.

Secret-key distillation usually comprises a step called *reconciliation*, whose purpose is to correct the transmission errors, and a step called *privacy amplification*, which wipes out Eve's information at the cost of a reduced key length. I shall briefly describe these two processes.

In the case of BB84, the reconciliation usually takes the form of an interactive error correction protocol. Alice and Bob alternatively disclose parities of subsets of their key elements. When they encounter a diverging parity, it means that there is an odd number of errors in the corresponding subset, hence at least one. Using a dichotomy, they can narrow down the error location and correct it. They repeat this process a sufficient number of times and the result is that Alice and Bob now share equal bits.

For secret-key distillation, all the communications are made over the public authenticated classical channel. Remember that Eve cannot intervene in the process but she may listen to exchanged messages, which in this case contain the exchanged parity bits. Therefore, the knowledge of Eve is now composed of $I_E + |M|$ bits, with $|M|$ the number of parity bits disclosed during the reconciliation.

To make the key secret, the idea behind privacy amplification is to exploit what Eve does not know about the key. Alice and Bob can calculate a function f of their key elements so as to spread Eve's partial ignorance over the entire result. Such a function (e.g., like a hashing function in classical cryptography) is chosen so that each of its output bits depends on most of, if not all, the input bits. An example of such a function consists of calculating the parity of random subsets of bits. Assume, for instance, that Eve perfectly knows the bit x_1 but does not know anything about the value of the bit x_2. If the function f outputs $x_1 + x_2 \bmod 2$, Eve has no clue on this output value since the two possibilities $x_1 + x_2 = 0 (\bmod 2)$ and $x_1 + x_2 = 1 (\bmod 2)$ are equally likely no matter what the value of x_1 is.

The price to pay for privacy amplification to work is that the output (secret) key must be smaller than the input (partially secret) key. The reduction in size is roughly equal to the number of bits known to Eve, and the resulting key size is thus $l - I_\text{E} - |M|$ bits. To maximize the key length and perhaps to avoid Eve knowing everything about the key (e.g., $l - I_\text{E} - |M| = 0$), it is important that the reconciliation discloses as little information as possible, just enough to make Alice and Bob able to correct all their errors.

Notice that errors on the quantum transmission are paid twice, roughly speaking, on the amount of produced secret key bits. First, errors should be attributed to eavesdropping and are counted towards I_E. Second, errors must be corrected, for which parity bits must be publicly disclosed and are counted towards $|M|$.

Finally, the secret key obtained after privacy amplification can be used by Alice and Bob for cryptographic purposes. In particular, they can use it to encrypt messages and thus create a secret channel.

1.1.4 Further reading

For more information, I should like to point out the paper by Bennett, Brassard and Ekert [12]. One can also find more technical information in the review paper by Gisin, Ribordy, Tittel and Zbinden [64].

1.2 Notation and conventions

Throughout this book, we use random variables. A *discrete random variable* X is a pair composed of a finite set $\mathcal{X}$ and a probability distribution on $\mathcal{X}$. The elements x of $\mathcal{X}$ are called *symbols*. The probability distribution is denoted as $P_X(x) = \Pr[X = x]$ for $x \in \mathcal{X}$ and of course verifies the relations $P_X(x) \geq 0$ and $\sum_x P_X(x) = 1$. We will use capital roman letters for random variables, the corresponding lower-case roman letters for the particular values (or symbols) that they can take, and the corresponding capital script letter for the sets over which they are defined.

The continuous random variables are defined similarly. A *continuous random variable* X is defined as an uncountable set $\mathcal{X}$ together with a probability density function $p_X(x)$ on $\mathcal{X}$.

The other important definitions are given along the way. For a list of the main symbols and abbreviations, please refer to the Appendix.

2
Classical cryptography

Following its etymology, the term *cryptography* denotes the techniques used to ensure the confidentiality of information in its storage or transmission. Besides confidentiality, cryptography also encompasses other important functions such as authentication, signature, non-repudiation or secret sharing, to name just a few.

The purpose of this section is to give a short introduction to classical cryptography, but only for areas that are relevant to quantum cryptography. For our purposes, we will only deal with confidentiality and authentication. The former is the most important function that quantum key distribution helps to achieve, while the latter is a requirement for quantum key distribution to work. Also, we will cover some topics on the security of classical cryptographic schemes so as to give some insight when comparing classical and quantum cryptography.

The study of cryptography ranges from the analysis of *cryptographic algorithms* (or *primitives*) to the design of a *solution* to a security problem. A cryptographic algorithm is a mathematical tool that provides a solution to a very specific problem and may be based on premises or on other primitives. A solution can be designed by combining primitives, and the requirements and functions of the primitives have to be combined properly.

We will first review the confidentiality provided by ciphers that use the same key for both encryption and decryption, hence called *secret-key ciphers*. Then, we will discuss authentication techniques. Finally, we will show some aspects of *public-key cryptography*.

2.1 Confidentiality and secret-key ciphers

To ensure the confidentiality of a transmission, the transmitter can use an algorithm (or a *cipher*) to transform (i.e., to *encrypt*) the data into apparent gibberish. Someone not knowing this algorithm will find the transmitted data meaningless. The receiver, however, knows how to invert the transformation and can read the data in their original form (i.e., he *decrypts* the transmitted data). The original data is called the *plaintext*, while the encrypted data is called the *ciphertext*.

Choosing a different algorithm for every transmission or for every transmitter–receiver pair is not very practical and suffers from fundamental drawbacks. Instead, modern cryptographers use a finite-sized set of cryptographic primitives and each transmitter–receiver pair ensures confidentiality of its communications by using a different *key*. A key is a parameter of a cipher, without which it is very difficult, if not impossible, to recover the original message. Among other things, Kerckhoffs' principle says that the security of a cryptosystem should not rely on the secrecy of the algorithm but only on the secrecy of the key [95]. In particular, it is assumed the adversary knows all the details of the cipher; only the key must be kept secret.

2.1.1 Perfect secrecy and the one-time pad

Before we describe any classical cipher, let us define the important concept of perfect secrecy. For this, we need to introduce the following notation. Let $P \in \mathcal{P}$ denote the random variable describing the plaintext messages (or symbols) that the legitimate users wish to transmit, and let $C \in \mathcal{C}$ be the corresponding ciphertext messages (or symbols). We here follow a statistical approach, as we model the plaintext data as a random variable.

A cipher is said to achieve *perfect secrecy* if for all $p \in \mathcal{P}$ and $c \in \mathcal{C}$,

$$\Pr[P = p | C = c] = \Pr[P = p],$$

or stated otherwise, if the knowledge of the ciphertext c does not change the statistical distribution of the plaintext messages.

Perfectly secret ciphers exist, and the most important and famous example is the *one-time pad*, created by combining the ideas of Vernam and Mauborgne in the years 1917–1918 [92, 166, 176]. For our description, the plaintext message $P = (p_i)$ is assumed to be formatted as a string of bits, but other alphabets can be used as well. The key $K = (k_i)$ consists of an independent string of bits, of length at least equal to the length of the plaintext message. The one-time pad consists in modulo-2 adding each plaintext bit p_i with each key bit k_i, namely $c_i = p_i \oplus k_i$, with $C = (c_i)$ the ciphertext. For instance, if the plaintext is 11001 and the key is 01101, then the ciphertext is $11001 \oplus 01101 = 10100$.

As its name suggests, the key of the one-time pad must be used only once.

Failing to do so will give the eavesdropper information on the plaintext. If p and p' are two plaintext bits encrypted with the same key bit k, then the modulo-2 addition of the two plaintext bits is equal to the modulo-2 addition of the two ciphertext bits, $p \oplus p' = c \oplus c'$. Although the attacker does not get the exact values of p and p', he is able to select two options (i.e., $(p, p') \in \{(0, c \oplus c'), (1, c \oplus c' \oplus 1)\}$) instead of four for the two plaintext bits. This gives not only information in the Shannon sense (see Chapter 3); it also allows practical cryptanalysis as long as the plaintext has enough redundancy [153].

To use the one-time pad correctly, Alice and Bob must create a secret key with as many bits as the length of the message (or messages) that they wish to send. This key K must be composed of independent unbiased random bits. With these requirements, the one-time pad achieves perfect secrecy, as shown by Shannon [161].

Perfect secrecy means that an eavesdropper, Eve, has no way to determine the plaintext if she has no idea about the content. This works even if the plaintext message is statistically biased, as would be the letters of a natural language. Even if the eavesdropper knows that Alice and Bob communicate, say, using Greek characters coded in Unicode, the ciphertext will give no extra information, and the best the eavesdropper can do is to guess any Greek text at random. In simple terms, acquiring the ciphertext does not help determining the plaintext at all as long as the key remains secret.

The one-time pad has often been used in military and diplomatic contexts. The key consisted of a long string of random zeroes and ones written on a long tape and transported in a diplomatic suitcase. As soon as it was used up, the tape would be destroyed.

In most of the ciphers used today, instead, the size of the key is much smaller than the size of the message to encrypt. This is indeed a good idea in practice. Establishing a secret key must be done carefully, as it must be kept secret from an eavesdropper. If Alice and Bob would need to secretly transmit a key as long as their secret message, they would probably use this effort to secretly transmit their message in the first place. Also, they may want to establish a very long key so as to foresee the traffic of all future messages. The only advantage of the diplomatic case is its tamper evidence. If someone got access to the key, Alice and Bob would make sure not to use it. Using a small key makes things easier in practice. Alice and Bob can establish a secret key once and for all, perhaps by meeting face-to-face, and then use it to encrypt as many secret messages as they wish.

While the use of a small key has undeniable practical advantages, it may not be used to achieve perfect secrecy. Indeed, Shannon showed that the key size must be at least as large as the message size for the cipher to be perfectly secure. Instead, ciphers usually used in classical cryptography are not perfectly secure and rely on the difficulty of solving certain problems, as we will explain later.

2.1.2 Stream ciphers

Stream ciphers are well-studied cryptographic primitives that somehow mimic the one-time pad, while using only a small secret key. For simplicity, we restrict ourselves to synchronous stream ciphers. The idea is to generate a long keystream $Z = (z_i)$, $i \in \{0 \ldots N-1\}$, from a secret key $K \in \mathbf{Z}_2^n$, where $N \gg n$, using a pseudo-random expansion function. The generated keystream can then be used to encrypt the plaintext, bit per bit, in a fashion identical to the one-time pad: $c_i = p_i \oplus z_i$. Although the size n of the key typically ranges between 128 and 256 bits, the keystream can be used to encrypt gigabytes of data.

For illustration purposes, let us describe a typical structure for a so-called synchronous stream cipher. Such a cipher can be described as a state machine, that is, a process that depends on a variable $S \in \mathbf{Z}_2^m$ called the *state*, which evolves during the production of the keystream. The state size m is usually greater than the key size.

To produce a keystream bit, the stream cipher outputs a function f of the key and of the state, $z_i = f(K, S(i))$, where $S(i)$ denotes the state at time i. Then, the state is updated $S(i+1) = g(K, S(i))$ using some function g before producing the next bit. At startup, the state S is initialized as a function of the key and some constant called the *initialization vector* (IV): $S(0) = h(K, \mathrm{IV})$. The IV can be publicly communicated and allows the generation of several different keystreams using the same secret key K.

There exist many synchronous stream ciphers based on *linear feedback shift registers* (LFSRs). In this case, the state S is structured as a sequence of bits,
$$S = (s_0, s_1, \ldots, s_{m-1}).$$
The state update function g consists in shifting all bits one position to the left, and the new value of the bit at the rightmost position is a linear function of the previous bits:
$$s'_i = s_{i+1} \text{ for } 0 \leq i \leq m-2,$$
$$s'_{m-1} = \sum_{j=0\ldots m-1} g_j s_j,$$

for some binary constants $g_0, \ldots, g_{m-1} \in \mathrm{GF}(2)$. The keystream can be obtained by extracting one bit from the state, for instance s_0, or by combining several bits from the state,

$$z_i = \sum_{j=0\ldots m-1} f_j s_j,$$

for some binary constants $f_0, \ldots, f_{m-1} \in \mathrm{GF}(2)$.

Stream ciphers based on LFSRs have been successfully cryptanalyzed and are not secure. The structure of this kind of stream cipher allows the eavesdropper to predict the whole keystream if some part of the keystream is known; this may happen if, for example, a part of the plaintext is accidently disclosed. Then, the keystream can be easily recovered from both the plaintext and ciphertext bits: $k_i = p_i \oplus c_i$. It is a simple matter of linear algebra to predict the next keystream bits and thus to be able to decrypt the rest of the ciphertext. Eve might even try to reverse the initialization function and recover the key K from the keystream. This way, she can predict other keystreams generated with the same key but with different IVs.

Even if the plaintext is not accidentally disclosed, it may sometimes be guessed by the eavesdropper. Eve might know that Alice transmits a zip file to Bob, which has a standard header, or she might assume that the transmitted message contains a letter that starts by some usual opening phrase, such as "Dear Bob,...". In all these cases, this gives Eve enough information to start decrypting the rest of the keystream.

Of course, Alice and Bob can combine several LFSRs, use non-linear operations or a more complex stream cipher structure to avoid this kind of problem. But these changes will not prevent the following fact: a stream cipher cannot be perfectly secure if the key is smaller than the keystream. For the one-time pad, the key and the keystream are the same; there are thus 2^N possible keystreams. For a stream cipher with a given IV, 2^n keys can "only" generate 2^n different keystreams, hence a much smaller number.

This does not mean that all stream ciphers are broken in practice. There exist plenty of stream ciphers that would require impractical resource levels for Eve to be able to predict the keystream or to recover the key. The simplest possible attack is the *exhaustive search*. The idea is to go through all the possible 2^n keys, decrypt the message, and stop when one gets a meaningful one. The exhaustive search quickly becomes impractical as n grows. Assuming a computer with a billion chips, each able to search through a billion different keys per second, the exhaustive search of a 96-bit key would take about 2500 years. For a 128-bit key, this would take four billion times longer. As eager as Eve might be to find the key, she will probably run out of patience before the end of the search.

Proofs of the security of stream ciphers currently do not exist. A cipher is considered secure when it has been significantly analyzed by the cryptographic community and no attack is found. Some stream ciphers have no currently known attacks better than exhaustive search. However, it does not mean that they do not exist: some clever advances in research may unveil new attacks. We will come back on this aspect later in Section 2.1.4.

2.1.3 Block ciphers

There is another important cryptographic primitive, called a *block cipher*, which is very often used in practice. A block cipher encrypts a block of b bits at a time using a key of n bits. Typical block sizes are $b = 64$, $b = 128$ and $b = 256$ bits, while secret key sizes typically range in $n \in \{56\ldots 256\}$.

For a given key $K \in \mathbf{Z}_2^n$, a block cipher F_K is a bijection in $\mathbf{Z}_2^b$. The encryption of the plaintext block $p \in \mathbf{Z}_2^b$ is denoted as

$$c = F_K(p).$$

The decryption of the ciphertext $c \in \mathbf{Z}_2^b$ is done using the inverse function F_K^{-1},

$$p = F_K^{-1}(c).$$

An important example of block cipher is the *Data Encryption Standard* (DES). The DES was developed by IBM in the early seventies under the solicitation of the US National Bureau of Standards. It was first published in 1975 and adopted as a standard in 1977 [60]. It encrypts blocks of $b = 64$ bits with a key of $n = 56$ bits.

In practice, the DES is no longer secure. Its *keyspace*, that is, the set of the 2^{56} possible keys, is too small and exhaustive search is within reach of current technologies. Also, attacks exist, which allow the key to be recovered faster than exhaustive search.

To overcome this, one often uses a variant called *Triple-DES*, which consists of the application of the DES three times with two independent keys,

$$\text{Triple-DES}_{K_1,K_2}(p) = \text{DES}_{K_1}(\text{DES}_{K_2}^{-1}(\text{DES}_{K_1}(p))).$$

In 1997, the US National Institute of Standards and Technology (NIST) made a formal call for new block ciphers to replace the DES. The new standard would be called the *Advanced Encryption Standard* (AES). Fifteen candidate algorithms, coming from various researchers throughout the world, were accepted by NIST. After an open evaluation process, the candidates were short-listed and five block ciphers remained in 1999: MARS, RC6, Rijndael, Serpent and Twofish. In 2000, the Belgian algorithm Rijndael [52, 132] was selected to become the AES. Created by Daemen and Rijmen, Rijndael was considered to have the best combination of security, performance and implementation flexibility among the candidates.

Rijndael is a block cipher of $b = 128$ bits, with a key size of $n = 128$, $n = 192$ or $n = 256$ bits. Other block sizes and key sizes exist but are not part of the AES specifications.

Block ciphers are usually constructed by iterating a simpler function, called a *round function*, a given number of times. Each iteration is called a *round* and the number of iterations is the *number of rounds*, denoted r. For the DES, the number of rounds is $r = 16$; for Rijndael, it is $r \in \{10, 12, 14\}$, depending on the chosen block and key lengths.

Let us denote a round function by $f_{K,i}$, where i is the round index. The block cipher function is thus obtained as

$$F_K = f_{K,r} \circ f_{K,r-1} \circ \cdots \circ f_{K,1}.$$

The round functions combine linear and non-linear operations so that the output bits depend in a complex way on the input and key bits. Within a few rounds, the influence of each input bit quickly propagates to all output bits. For instance, in the DES, every intermediate result bit is a function of every plaintext bit after only five rounds. The ingredients of the DES include non-linear functions, called S-boxes, that map 6 bits onto 4. The plaintext bits are permuted, XOR-ed and combined with key bits. In AES, the plaintext bytes are processed using both linear operations in $GF(2^8)$ and non-linear 8-bit-to-8-bit S-boxes.

I will not describe the block ciphers any further, as that would extend beyond the scope of this book. The description of the internal structure of the DES and the AES can be found in most books on classical cryptography, e.g., [127, 153, 170]. Nevertheless, I will take advantage of this macroscopic description in terms of round functions to introduce two important classes of cryptographic attacks, namely the *differential* and *linear cryptanalysis*.

Differential cryptanalysis

Differential cryptanalysis was introduced in 1990 by Biham and Shamir [20, 21, 22]. It is a class of attacks that uses *chosen plaintexts*, that is, we assume the attacker has the freedom of choosing data blocks to encrypt and of acquiring many plaintext-ciphertext pairs $\{(P_i, C_i = F_K(P_i))\}$. From these pairs, the purpose of the attack is to recover the unknown key K.

To fix the ideas, we may think that Alice and Bob encrypt their data using some secure device, which allows the key to be used in the block cipher but which does not allow anyone to read out the key; i.e., it is tamper-resistant. Eve, who regularly records Alice's and Bob's encrypted traffic, wants to recover the key by stealing the encryption device so as to be able to decrypt the past messages.

Differential cryptanalysis looks at what happens if we encrypt pairs of plaintext messages with a constant difference. Let ΔP be a b-bit constant. Eve chooses random plaintexts P_i and, for each, encrypts both P_i and $P_i^* = P_i \oplus \Delta P$ so as to obtain the ciphertexts $C_i = F_K(P_i)$ and $C_i^* = F_K(P_i^*)$. Note that the complex non-linear structure of a block cipher will not yield a constant output difference, that is, $\Delta C_i = C_i \oplus C_i^*$ is not constant.

Note that the attack can also be mounted without Eve choosing the plaintext messages. If she collects a large amount of known plaintext and ciphertext messages, she can also extract from them the pairs that have the desired difference. The attack is then less efficient because part of the data has to be thrown away. In this case, the attack is said to be a *known plaintext* attack.

We can analyze the output difference of a single round function. For a given input difference ΔX, the output difference is typically biased. Among randomly-chosen plaintext pairs $(X_i, X_i \oplus \Delta X)$, some output differences $\Delta Y_i = f_K(X_i) \oplus f_K(X_i \oplus \Delta X)$ may occur more often than others.

To perform differential cryptanalysis, we use difference biases over $r - 1$ rounds. More specifically, Eve chooses a plaintext difference ΔP so that it gives the difference ΔX with high probability after all but the last round. We call this a *differential*. Here, the probability is taken over the possible plaintext P_i and key K values. The difference ΔP is carefully chosen by Eve so that it gives the highest output bias after $r - 1$ rounds, and ΔX is the corresponding output difference.

In practice, one looks at one-round differences. For a multi-round differential, the attacker analyzes how the input difference ΔP propagates between each round and becomes ΔX at the end. A particular propagation pattern is called a *differential trail*. Note that several differential trails can have the specified input and output differences, hence contributing to the same differential.

One can choose a differential over only $r - 1$ rounds so that the last round can give us information on the key. Note that in practice, other attacks may work slightly differently, but my goal is to give the reader an intuitive idea.

Eve will thus generate a large amount of $(P_i, P_i \oplus \Delta P)$ pairs and get the corresponding ciphertexts (C_i, C_i^*) pairs from the secure device. Since she does not know the key, she does not know the value of the input of the last round, so she does not know when the difference ΔX occurs at the input of the last round. For a given pair (C_i, C_i^*), she will determine which key values satisfy the difference ΔX. She thus guesses a key value $\bar{K}$ and inverts the last round so as to determine whether

$$\Delta X \stackrel{?}{=} f^{-1}_{\bar{K},r}(C_i) \oplus f^{-1}_{\bar{K},r}(C_i^*). \tag{2.1}$$

For each key, she counts the number of times this condition is verified among all the plaintext pairs. Finally, she picks up the key for which the condition Eq. (2.1) was the most often verified. The correct key makes the correct bias emerge, while other keys produce unrelated values for which, in general, no particular bias appears. Hence, the correct key verifies Eq. (2.1) more often than the others.

At this point, it looks as if Eve has to go through all the possible keys $\bar{K}$, which would then make the attack inefficient, with the same complexity as an exhaustive search. The trick is that she does not have to guess all the key bits. The round function f is simple enough so that she can guess only a subset of the bits at a time. By working on independent subsets at a time, the process becomes much less complex than an exhaustive search.

How many plaintext blocks are necessary for the attack to be successful? For a given differential, the final bias is essentially determined by the probability ϵ that the difference ΔX occurs at the input of the last round given that the difference ΔP was used at the input of the first round. For the bias to emerge, the number of pairs needed is essentially ϵ^{-1}. Of course, Eve may try to find the differential with the highest probability, hence minimizing her efforts. Thus, to protect against differential analysis, the design of a block cipher must ensure that all differentials have roughly the same probability, 2^{-b}, and this is the case for all keys.

For the DES, the best known differential cryptanalysis requires 2^{47} plaintext blocks, hence requires a complexity of 2^{47} DES encryption when the plaintexts are chosen. This is faster than exhaustive search. In a more realistic scenario where the plaintexts are not chosen but only known, the complexity grows to 2^{55}, which is essentially the exhaustive search complexity. It is worth noting that the DES was designed before differential cryptanalysis was found by the research community, although it was probably secretly known to the NSA, who was involved in the design at that time.

The AES, on its side, was designed with resistance against differential cryptanalysis in mind. As a design criteria, the authors proved that no 8-round differential trails exist with a probability above 2^{-300}.

Linear cryptanalysis

Linear cryptanalysis was introduced in 1993 by Matsui [115, 116]. It relies on linear approximations that may exist between subsets of bits in the plaintext, ciphertext and key. In contrast to differential cryptanalysis, linear cryptanalysis is primarily a known plaintext attack.

The idea of the attack is the following. Let us take the modulo-2 addition of some plaintext bits, ciphertext bits and key bits, e.g.,

$$\Sigma = \sum_i \lambda_i^{\text{p}} p_i + \sum_i \lambda_i^{\text{c}} c_i + \sum_i \lambda_i^{\text{k}} k_i, \qquad (2.2)$$

where the subscripts indicate the positions within the blocks and key, and where $\lambda_i^{\text{p,c,k}} \in \text{GF}(2)$ are binary constants that select which bits are used in the sum. If the plaintext, ciphertext and key bits are completely random, Σ is equal to zero or one, each with a probability 1/2. However, in a block cipher, the ciphertext is deterministically computed from the plaintext and the key. So, there may be expressions of the form Eq. (2.2) that are biased towards zero or one.

Let us assume that Eve finds a correlation of the form Eq. (2.2) so that Σ is biased towards, say, zero. Given a large number of plaintext and ciphertext blocks, the attacker can calculate $\Sigma^{\text{p,c}} = \sum_i \lambda_i^{\text{p}} p_i + \sum_i \lambda_i^{\text{c}} c_i$ and count the number of times this expression is zero or one. If the majority goes to zero, this means that $\Sigma^{\text{k}} = \sum_i \lambda_i^{\text{k}} k_i = 0$; otherwise, $\Sigma^{\text{k}} = 1$. This gives one bit of information on the key. To recover the whole key, several correlations must be used.

The stronger the bias, the quicker the convergence of $\Sigma^{\text{p,c}}$. Finding a correlation with a strong bias is not obvious. As for differential cryptanalysis, the attacker can analyze the linear correlations within a single round of the block cipher and then combine these relations to the whole block cipher. The propagation of correlations within rounds is called a *linear trail*. Several linear trails can interfere to make a correlation.

What is the complexity of the attack? Let $\epsilon = 2(\Pr[\Sigma = 0] - 1/2)$ be the correlation coefficient of some linear approximation. Note that $\epsilon > 0$ if Σ is biased towards zero and $\epsilon < 0$ if it is biased towards one. Then, the number of necessary plaintext and ciphertext blocks is essentially ϵ^{-2}.

For the DES, the fastest known linear attack requires a complexity of 2^{43} DES operations and a pool of 2^{43} plaintext and ciphertext blocks. A variant requiring 2^{38} blocks also exist, but its complexity is 2^{50}. As I already mentioned, the DES is no longer considered secure, and the existence of known plaintext attacks with complexity much less than an exhaustive key search confirms this.

Like differential cryptanalysis, linear cryptanalysis was taken into account by the designers of the AES. As a design criteria, they proved that no 8-round linear trails exist with a correlation coefficient above 2^{-150}.

Other attacks

There are many other families of attack beyond linear and differential cryptanalysis. What is important to know is that there are attacks for which Eve does not need to have many pairs of corresponding plaintext and ciphertext blocks. Some attacks, e.g., algebraic attacks, try to model the block cipher as a huge system of binary equations with the key bits as unknowns. Provided that such attacks work on a given block cipher, only one pair of plaintext and ciphertext blocks, or a very small number of them, is necessary.

2.1.4 Security by peer review

We have seen that, besides the one-time pad, none of the ciphers used in practice today offers perfect secrecy. If Eve has enough computing power, she may be able to break a given cipher, that is, to decrypt encrypted data without knowledge of the key. In contrast to perfect secrecy, we say that a cipher is *computationally secure* if its security relies on the computational difficulty to break it.

Hopefully, as discussed above, an exhaustive search over a 128-bit keyspace would require an unreasonable computing power running for an unreasonable time. Furthermore, if technological advances increase the computer power faster than expected, cipher designers can simply increase the key size to keep a good security margin. So, computational security may not be a problem in practice.

But is exhaustive search the only option to break the cipher? Of course, there may be shortcut attacks. The DES, for instance, is breakable using linear cryptanalysis with a much lower complexity than exhaustive search. The AES, on its side, is designed to be resistant against differential and linear cryptanalysis and a number of other families of attacks that are beyond the scope of this book. But that does not exclude the existence of other, unknown, attacks. There are no proofs of a (non-trivial) lower-bound for the complexity of breaking the AES.

How would one consider a particular cipher secure, or at least, secure enough for a given application? Peer review is the usually admitted process. The research community believes that a cipher is secure if enough experts analyzed it and no one found any shortcut attacks. The notion of "enough" is of course subjective. In the particular case of the AES, the algorithm has been reviewed by hundreds of cryptanalysts throughout the world for almost ten years now. At this time of writing, no shortcut attack has been found and the majority of the cryptographic community believes that the AES is secure – see, for instance, the proceedings of the Fourth Conference on AES [84].

There have been many proposals for new ciphers, some of which got more attention than others. Many have been broken, many others are still unbroken. New designs usually take into account the existing attacks, building upon the experience acquired for other ciphers. The security of ciphers increases as the power of cryptanalysis increases; what is considered secure today may not be tomorrow. The AES is likely to stand for as many years as it has already existed. But surprises may be ahead; no one knows.

2.2 Secret-key authentication

The second important function of cryptography we need to introduce is *authentication*. Authentication is the set of techniques aimed at verifying that a transmitted message arrives unmodified at the recipient's station; this is also sometimes called message integrity. Messages modified or created by any person other than the legitimate sender (i.e., an adversary) can be detected and discarded.

Authentication is important in a wide number of applications. For Internet shopping, for instance, the store wants to make sure that the customer is genuine, so that the money comes from a real bank or credit card account. And, vice versa, one does not want to buy hollow products from a fake online store. In this example, authentication comes before confidentiality. Except for specific products, the customer does not really care if an eavesdropper can determine which books s/he is buying.

In the scope of quantum cryptography, authentication is an essential ingredient. We have seen that eavesdropping on the quantum channel implies an increase in transmission errors. At some point, Alice and Bob have to sample their transmission so as to measure transmission errors and detect eavesdropping. When doing so, Eve might intercept the exchanged messages and make the legitimate parties believe there are no transmission errors, hence hindering the eavesdropping detection. To prevent this, Alice and Bob have to make sure they talk over an authenticated channel.

By protecting also the message content, authentication prevents *man-in-the-middle* attacks (or rather *woman-in-the-middle* in the case of Eve). As an example, assume that a customer is talking to a legitimate online bookstore. Upon check-out, he sends the address to which the book must be sent. Eve, who is monitoring the communications, changes the address and puts her own instead. If the messages are not authenticated, the legitimate parties do not realize this, and Eve receives the book at the customer's expense.

In the light of the example above, it is essential that authentication depends on the message content. Authentication usually comes down to adding some form of redundancy to the original message. Any modification to the message would most likely break the expected redundancy. The information added to the original message is called a *message authentication code* (MAC).

With secret-key authentication, the same key is used to generate and to verify the MAC. We must thus assume that Alice and Bob share a secret key K. Before sending a message, Alice attaches a MAC, which she calculates as a function of both the message and the key. On his side, Bob recalculates the MAC from the received message and compares it to the attached MAC. Provided that the shared key is not compromised, he can be confident that the message is legitimate when both MACs match.

There are two kinds of attacks against an authentication scheme. The first kind is *impersonation*: after seeing t messages m_1 to m_t, Eve creates a message m'_{t+1} that she wants to be accepted by the receiver as legitimate. The second kind is *substitution*: after seeing t messages m_1 to m_t, Eve modifies a message m_t and replaces it with her own message m'_t that she wants to be accepted by the receiver as legitimate.

2.2.1 Authentication using a block cipher

Message authentication codes can be built upon block ciphers. For instance, let us illustrate this with the cipher block chaining MAC construction.

Let m be the message to authenticate, which we assume is composed of β blocks of length b each. Note that some padding may be used for the message length to reach a multiple of b bits. The individual blocks are denoted as m_i for $1 \leq i \leq \beta$. Let F_K be a block cipher with key K.

The first block is encrypted, $a_1 = F_K(m_1)$. Then, the second block is bitwise modulo-2 added to a_1 before it is encrypted, that is, $a_2 = F_K(m_2 \oplus a_1)$. For each of the next blocks, one computes $a_i = F_K(m_i \oplus a_{i-1})$. Finally, the value of the last encryption MAC $= a_\beta$ is used as a MAC. Note that, to make the MAC value depend on the message size β, the last block may undergo some additional operations.

The security of this scheme depends on the strength of the underlying block cipher. If the block cipher is secure, this construction gives a strong

MAC in the following sense. First, it is difficult to calculate a correct MAC on a chosen message if the key is unknown (resistance against impersonation). Then, for a given (m, MAC) pair, it is difficult to find another valid $(m' \neq m, \text{MAC}')$ pair for a different message (resistance against substitution).

2.2.2 Intrinsic security of authentication

An important fact to remember about authentication is that it can be achieved without any computational assumptions. Message authentication codes (MACs) based on computationally-secure cryptographic primitives, such as those based on block ciphers, are most often used in practice. In the scope of quantum cryptography, however, the idea is to step away from computational assumptions, so in this section we consider secure authentication without computational assumptions.

As long as Alice and Bob share some initial secret key K, the theory guarantees that an authentication protocol exists such that the success of an attacker, Eve, being able to forge messages does not depend on her computational resources. One example of protocol is based on MACs calculated from (keyed) universal families of hash functions [169, 182]. I will give an example of construction in Section 5.1.1.

Note that the key bits are consumed each time a MAC is computed. As for the one-time pad, new key bits are, in general, necessary for each new message to authenticate.

Another important fact, however, is that authentication is never going to be perfect. In contrast to confidentiality, which can be made perfect (e.g., with the one-time pad, see Section 2.1.1), there is always a non-zero probability of success from an active – and probably lucky – attacker. The reason for this is that Eve can choose a key K at random, compute a MAC with this key and see if her message gets accepted by Bob. The probability of this happening may be very small, but once she knows she has the correct key, she can go on with it and her messages become indistinguishable from Alice's.

More subtle attacks can be mounted, combining impersonation and substitution. As explained in [121], no authentication scheme can prevent Eve from being successful with a success probability lower than $|\mathcal{K}|^{-1/2}$, with $|\mathcal{K}|$ the size of the keyspace. For a given requirement on the success probability p_{Auth}, the key must thus contain at least $-2 \log p_{\text{Auth}}$ bits.

2.3 Public-key cryptography

For the secret-key ciphers and authentication schemes that we saw in the previous sections, the same secret key is shared by both parties. For confidentiality, the same key serves both encryption and decryption purposes; for authentication, it is used both to generate and to verify a MAC.

Public-key ciphers are quite different from their secret-key counterparts. Before a public-key cipher is used, the user has to create a public key and private key pair. These two keys are related in such a way that, for confidentiality purposes, the public key can be used to encrypt data, while the ciphertext can be decrypted only with the knowledge of the corresponding private key. The user can publish his public key in a directory so that anyone can send him confidential messages, as he is the only person having the corresponding private key.

2.3.1 Confidentiality with RSA

As an example, let us briefly describe the well-known public-key cipher called RSA from the name of its inventors Rivest, Shamir and Adleman [150].

With RSA, Bob starts by selecting two large prime numbers p and q, which he multiplies so as to create a number n. By "large", we mean that they have, say, 500 or more bits (150 or more decimal digits); the length of n expressed in binary is usually above 1000 bits.

Bob also chooses a number $3 \leq e < n$ such that e and n have no common factor. He calculates d such that $ed = 1 \;(\mathrm{mod}(p-1)(q-1))$. Once this is done, he may discard the prime numbers p and q. He publishes the pair (e, n) as his public key and secretly keeps the pair (d, n) as his private key.

Encryption goes as follows. For a plaintext message $m \in \mathbf{Z}_n$, Alice calculates its e-th power and reduces the result modulo n; or simply stated, she evaluates $c = m^e \bmod n$.

Upon reception, Bob can decrypt the message using his private key by computing $c^d \bmod n$. The fact that $ed = 1 \;(\mathrm{mod}(p-1)(q-1))$ and the properties of modular exponentiation imply that

$$c^d \bmod n = m^{ed} \bmod n = m. \tag{2.3}$$

The creation of the modulus $n = pq$ is done privately by Bob. If the prime factors are known, then anyone can easily find the private exponent d from e just as Bob did during the creation of the key. So, an attacker could factor n, which is part of the public key, and derive the private key. This is why Bob chooses large prime factors; multiplying these two numbers is easy, but factoring n to find them is much harder.

RSA is a computationally-secure cipher, as it relies on the difficulty to factor large numbers. The fastest known algorithms require a time that grows faster than any polynomial in $\log n$, the number of digits of the modulus to factor. So, factoring does indeed seem like a difficult problem. However, there may be faster algorithms that are yet to be discovered and one currently does not know the minimal complexity factoring.

Furthermore, factoring n may not be the only way to break RSA. If the goal is to decrypt a message without the knowledge of the private key, there may be clever tricks to do so without factoring n. There is currently no proof that RSA and factoring are computationally equivalent, although many believe so.

2.3.2 Signatures with RSA

For confidentiality, the exponent e is used to encrypt, while the exponent d is used to decrypt. The symmetry of RSA allows e and d to be interchanged. If we do so, we obtain a *signature* scheme.

Assume that Bob "encrypts" a message m with his *private* key, that is, he computes $s = m^d \bmod n$. Like Eq. (2.3), the properties of e and d imply that $s^e \bmod n = m$. This means that anyone knowing Bob's public key (e, n) can decrypt the message; the "encrypted" message s does not provide any confidentiality. However, there is a link between e and d that is uniquely known to the creator of the key (i.e., Bob). If anyone can revert the encryption, it means that only Bob could have done it. Hence, this provides authentication on the message m.

Public-key authentication is fundamentally different from secret-key authentication and it is usually called a signature scheme, as anyone can verify the authentication, not only the recipient.

In practice, signatures with RSA are more involved. For instance, the message m can be longer than the modulus n and the full message is usually mapped to a shorter message by applying a *message digest function* (or *cryptographic hash function*). Furthermore, the symmetry of RSA would allow an active attacker to be able to create signatures on fake messages by combining the signatures of past messages. Special techniques are used to prevent this. However, the basic idea remains: the private key is used to generate the signature while the public key is used to verify it.

2.3.3 Diffie–Hellman key agreement

The Diffie–Hellman key agreement protocol [54] provides each pair of users, Alice and Bob, with a common secret key by way of public-key techniques. As for RSA, each user has a private and a public key. The result of the key agreement is a secret key, unique to a given pair of users, which can be used for confidentiality or authentication purposes using secret-key techniques, e.g., block ciphers.

Note that this combination is not unique to the Diffie–Hellman protocol. RSA encryption could be used as well. Alice would select a random secret key K, encrypt it with Bob's RSA public key and send it to him. Bob would decrypt the key K and both parties would then be able to use secret-key technique.

This combination is interesting in practice because public-key techniques are, in general, slower than their secret-key counterparts. If Alice wishes to send Bob a large message, it is much more efficient first to establish a secret key and then to use a computationally efficient secret-key cipher than to encrypt the whole message with a public-key primitive.

Let us now describe Diffie–Hellman more explicitly. First, a prime number p is chosen, together with a generator g of Z_p^*. We say that g is a generator of Z_p^* if $\{g^i \bmod p : 0 \leq i \leq p-1\}$ takes all the possible values in Z_p^*. These parameters are public and are common to all the users of the scheme.

To set up the protocol, Alice (or Bob) randomly chooses a secret key $1 \leq a \leq p-2$ (or $1 \leq b \leq p-2$) and computes $g^a \bmod p$ (or $g^b \bmod p$). They publish $g^a \bmod p$ and $g^b \bmod p$ as their public keys and keep the exponents as their private keys.

To establish a common secret key, Alice and Bob take each other's public key and raise it to the powers a and b, respectively. Specifically, Alice calculates $K = (g^b)^a \bmod p$, while Bob calculates $K = (g^a)^b \bmod p$. Since the exponents commute, they obtain the same value K, which is the common secret key.

2.3.4 Public-key versus secret-key cryptography

I shall now highlight some of the important differences between public-key and secret-key cryptography.

The function offered by public-key cryptography (PKC) is fairly different from that of the secret-key cryptography (SKC). With PKC, anyone can encrypt to a single recipient, Bob, while SKC is essentially intended for one-to-one communications. With SKC, Alice and Bob have interchangeable roles since they share the same key. In PKC, however, the recipient Bob is attached to its private key, while the sender can be anyone having access to the directory of public keys.

The same idea applies to authentication. With secret-key authentication, nobody else other than Alice and Bob can verify each other's MACs. Since they have the same key, they can both generate and verify MACs. From the point of view of a third party to which Alice and Bob would give the key, nothing distinguishes a message authenticated by Alice or by Bob. With (public-key) signatures, however, only the owner of the private key can generate a valid signature and anyone can verify it.

In a network of n people, SKC would impose everyone to establish a secret key with everyone else, hence resulting in a potential number $n(n-1)/2$ of different keys. With PKC, only n pairs of private and public keys are needed. For this reason, PKC is usually considered to be more practical than SKC.

2.3.5 Public key infrastructure

It must be stressed that using someone's public key is more complex than a simple lookup in the directory. The identity of the intended recipient, Bob, and his public key must be linked together, otherwise the enemy would be offered an easy path for a man-in-the-middle attack.

Imagine that Alice wants to send a message to Bob using his public key. An attacker, Eve, sends to Alice her own key, pretending she is Bob. Alice encrypts her message, intended to Bob, using Eve's public key. The eavesdropper is thus able to decrypt the message and to forward it to Bob using his public key. Eve could even do the same in the other direction, reading Alice's and Bob's messages while going unnoticed.

In *public-key infrastructure* (PKI), this problem is solved by associating a key with a certificate. This certificate provides a signature, made by a higher authority, namely, a certification authority (CA), that shows that a given public key belongs to a given person. Before publishing his key, Bob goes to his local CA offices, gives his public key and shows evidence of his identity. After verification, the CA signs Bob's key and gives him the certificate.

This solution can be made hierarchical, as there might be local CAs, whose keys are signed by some higher-level CA. Ultimately, one has to trust the top CA's public key to be able to verify certificates. The top CA's public key is usually given in the setup of the system. For instance, a web browser usually comes with built-in CA public keys. It can then verify the certificates of other keys and establish a secure connection with any web site whose public key is certified.

2.4 Conclusion

In this chapter, I showed some aspects of classical cryptography that are relevant to the context of quantum cryptography. In particular, I described the important concept of perfect secrecy and discussed the assumptions on which the security of practical cryptographic primitives are based. Finally, I highlighted some of the differences between secret-key and public-key cryptography.

More information on cryptography can be found in the books by Schneier [153], by Stinson [170] and by Menezes, van Oorschot and Vanstone [127].

3
Information theory

Founded by Shannon, information theory deals with the fundamental principles of communication. The two most important questions answered by this theory are how much we can compress a given data source and how much data we can transmit in a given communication channel.

Information theory is essentially statistically minded. Data sources are modeled as random processes, and transmission channels are also modeled in probabilistic terms. The theory does not deal with the content of information – it deals with the frequency at which symbols (letters, figures, etc.) appear or are processed but not their meaning. A statistical model is not the only option. Non-statistical theories also exist (e.g., Kolmogorov complexity). However, in this section and throughout this book, we will only use the statistical tools.

Information theory is of central importance in quantum cryptography. It may be used to model the transmission of the key elements from Alice to Bob. Note that what may happen on the quantum channel is better described using quantum information theory – see Chapter 4. Yet, the key elements chosen by Alice and those obtained by Bob after measurement are classical values so, for instance, the transmission errors can accurately be modeled using classical information theory. Reconciliation, in particular, requires classical information-theoretic techniques.

3.1 Source coding

Source coding is the first problem that information theory addresses. Assume that a source emits symbols x_i from an alphabet $\mathcal{X}$ and that it can be modeled by the random variable X on $\mathcal{X}$. For instance, the source can be the temperature measured by some meteorological station at regular intervals or the traffic on a network connection. The emitter wishes to send the symbols produced by the source to a recipient via a reliable transmission channel that transmits bits. For economical or timing reasons, the emitter wishes to compress the source, that is, to encode the source with the least number of bits; this is the source coding problem.

For source coding and other information-theoretic problems, the Shannon entropy is a central concept that we need to define before going any further.

The *Shannon entropy* (or *entropy* for short) of a discrete random variable X is denoted by $H(X)$ and is defined as

$$H(X) = -\sum_x P_X(x) \log P_X(x).$$

By convention, all logarithms are in base 2, unless otherwise stated. Note that the entropy is always positive, $H(X) \geq 0$. An important special case is the entropy of a binary random variable with distribution $\{p, 1-p\}$ for $0 \leq p \leq 1$, which is denoted

$$h(p) = -p \log p - (1-p) \log(1-p).$$

The Shannon entropy is a property of the distribution $P_X(x)$ but not of the symbol set $\mathcal{X}$.

In source coding, the entropy of a random variable tells us precisely about its compressibility. Assuming a finite range $\mathcal{X}$, a random variable X could be encoded using $\lceil \log |\mathcal{X}| \rceil$ bits per symbol but this is, of course, not optimal if some symbols are more frequent than others. One can imagine encoding more frequent symbols using fewer bits and vice versa. We call the *rate R* the average number of bits per symbol.

The approach of encoding more frequent symbols using fewer bits is taken for instance by Huffman coding – see Section 3.1.2 below. The properties of Huffman coding guarantee that the rate R satisfies

$$H(X) \leq R < H(X) + 1.$$

Thus, up to a one bit per symbol variation, the Shannon entropy tells us the rate of the Huffman coding. In fact, Shannon showed that it is not possible to compress a source with a rate lower than its entropy [160]; the inequality $H(X) \leq R$ applies to all source codes.

To improve the rate, we can compress d symbols jointly. Assuming that the symbols are produced independently of one another, it is easy to prove that $H(X^{(d)}) = dH(X)$, where $X^{(d)}$ is the random variable obtained by concatenating d independent instances of X. Then, we Huffman-encode $X^{(d)}$ and obtain a (per symbol) rate R that satisfies $H(X^{(d)}) \leq dR < H(X^{(d)})+1$,

or equivalently
$$H(X) \leq R < H(X) + d^{-1}.$$
By increasing d, we can be convinced that the compression rate can approach Shannon's entropy as closely as is desired.

3.1.1 Properties of source codes

Let me now describe in more detail the family of codes that one can use for source coding.

A *source code* α for a random variable X is a mapping from $\mathcal{X}$ to the codewords $\{0,1\}^*$, the set of finite binary strings – the star denotes the concatenation of zero, one or any finite number of symbols. We consider only *binary* codes. The codeword associated with $x \in \mathcal{X}$ is written as $\alpha(x)$.

The *length* of a codeword $\alpha(x)$ is, of course, the number of bits that compose it and is denoted as $|\alpha(x)|$. For a source code, we define the *average length* as
$$L(\alpha) = \sum_{x \in \mathcal{X}} P_X(x) |\alpha(x)|.$$

A code α is *non-singular* if every $x \in \mathcal{X}$ is encoded into a different codeword $\alpha(x)$. This ensures that we can decode x given its codeword $\alpha(x)$.

We usually want to encode many elements of $\mathcal{X}$ by concatenating their binary codes. This defines a codeword for all strings $\bar{x} = x_1 x_2 \ldots x_{|\bar{x}|} \in \mathcal{X}^*$ into $\{0,1\}^*$: $\alpha(\bar{x}) = \alpha(x_1)\alpha(x_2)\ldots\alpha(x_{|\bar{x}|})$, where the composition law is the concatenation. This new code should be such that we can recover the string in $\mathcal{X}^*$. We say that a code is *uniquely decodable* if for all strings $\bar{x} \in \mathcal{X}^*$, the resulting codeword $\alpha(\bar{x})$ is different.

Being uniquely decodable does not mean that the binary string is easy to parse and to cut into individual codewords. To describe codes that are easy to decode, we need to introduce a few definitions. The binary string s_1 is said to be a *prefix* of s_2 if the $|s_1|$ first bits of s_2 are equal to those of s_1. We say that s_1 is a *proper prefix* of s_2 if s_1 is a prefix of s_2 and $s_1 \neq s_2$. A code is said to be *instantaneous* or *prefix-free* if no codeword is a prefix of another codeword.

As an example, let us take the alphabet $\mathcal{X} = \{a, b, c\}$ and two different codes:
$$\alpha_0: (\alpha_0(a), \alpha_0(b), \alpha_0(c)) = (1, 11, 111),$$
$$\alpha_1: (\alpha_1(a), \alpha_1(b), \alpha_1(c)) = (1, 10, 100),$$
$$\alpha_2: (\alpha_2(a), \alpha_2(b), \alpha_2(c)) = (1, 01, 001).$$

First, the code α_0 is not uniquely decodable because, e.g., $\alpha_0(ac) = \alpha_0(bb) = 1111$. Then, the code α_1 is uniquely decodable but not prefix-free. The prefix condition fails because, e.g., $\alpha_1(a)$ is a prefix of the other two codewords. To decode one symbol from α_1, notice that a bit 1 can only appear at the beginning of a codeword. The decoder has to wait for the next codeword and count the number of 0 before the next 1 or before the end of the string. E.g., 110010100 is unambiguously parsed as $1, 100, 10, 100$ and decodes as $acbc$. Finally, the code α_2 is uniquely decodable and prefix-free. The decoding of a symbol does not require to wait for the next codeword. If the codeword starts with 1, it decodes as a, otherwise it is either b or c. Then b and c can be distinguished by reading the second bit of the codeword.

3.1.2 Huffman coding

Huffman codes were discovered in 1952 [85] and are an example of prefix-free codes. They enjoy the property of being optimal, that is, their average length L is minimal and verifies the equation $H(X) \leq L < H(X) + 1$. The construction is fairly easy and is described below.

- First, all the symbols of $\mathcal{X}$ are assigned to the empty codeword.
- Then, consider the two symbols with the lowest associated probabilities, which we note by y and z. We assign the codewords $\alpha(y) = 0$ and $\alpha(z) = 1$ to y and z. The symbols y and z are discarded from $\mathcal{X}$ and replaced by a new meta-symbol $y' = \{y, z\}$ with associated probability $P_X(y') = P_X(y) + P_X(z)$.
- We again consider the two symbols (or meta-symbols) x and y with the lowest probabilities. We prefix x and y by 0 and 1, that is,

$$\alpha(x) \leftarrow 0\alpha(x),$$
$$\alpha(y) \leftarrow 1\alpha(y).$$

If x (or y) is a meta symbol, then all the symbols contained in x (or y) are prefixed with 0 (or 1). The (meta-)symbols x and y are discarded and replaced by the meta-symbol $x' = x \cup y$ with $P_X(y') = P_X(y) + P_X(z)$.
- The previous step is repeated until $\mathcal{X}$ contains only one (meta-)symbol.

As an example, consider the alphabet $\mathcal{X} = \{a, b, c, d\}$ with probabilities

$$(P_X(a), P_X(b), P_X(c), P_X(d)) = (0.1, 0.1, 0.3, 0.5).$$

For the first step, a and b are assigned to $\alpha(a) = 0$ and $\alpha(b) = 1$ and replaced by a' with probability $P_X(a') = 0.2$. As the second step, a' and c have the lowest probabilities among the remaining symbols and receive a prefix 0 and 1, respectively, so that $\alpha(a) = 00$, $\alpha(b) = 01$ and $\alpha(c) = 1$. They are replaced by a'' with probability $P_X(a'') = 0.5$. Finally, a'' and d are prefixed with 0 and 1, so that $\alpha(a) = 000$, $\alpha(b) = 001$, $\alpha(c) = 01$ and $\alpha(d) = 1$.

3.1.3 Arithmetic coding

As an alternative to Huffman coding, arithmetic coding is an efficient candidate. This coding scheme is particularly well suited to the encoding of long streams of symbols. Unlike the source coding schemes we have seen so far, arithmetic coding is defined on strings $\bar{x} \in \mathcal{X}^*$, but is not defined on individual symbols $x \in \mathcal{X}$. A real number $r \in \mathbf{R}$ between 0 and 1 is associated with every string $\bar{x} = x_1 x_2 \ldots x_{|\bar{x}|} \in \mathcal{X}^*$. The codeword consists of the most significant bits of r expressed in binary.

To proceed with arithmetic coding, we have to impose an order to the symbols in $\mathcal{X}$. The ordering is arbitrarily chosen as a convention and does not influence the properties of the encoding. Let $\mathcal{X} = \{s_1, s_2, \ldots, s_n\}$ with $s_1 < s_2 < \cdots < s_n$.

The real number $r = r_1 = \alpha(\bar{x})$ is chosen so that

$$\Pr[X < s_j] \leq r_1 \leq \Pr[X \leq s_j]$$

iff the first symbol x_1 of $\bar{x}$ is s_j. Then, let

$$r_2 = \frac{r_1 - \Pr[X < x_1]}{\Pr[X \leq x_1] - \Pr[X < x_1]}$$

be the rescaled value of r_1 within the interval $[\Pr[X < x_1] \ldots \Pr[X \leq x_1]]$. The value r_2 is determined by the second symbol x_2 of $\bar{x}$: $\Pr[X < x_2] \leq r_2 \leq \Pr[X \leq x_2]$. And so on;

$$\Pr[X < x_i] \leq r_i \leq \Pr[X \leq x_i]$$

with

$$r_i = \frac{r_{i-1} - \Pr[X < x_{i-1}]}{\Pr[X \leq x_{i-1}] - \Pr[X < x_{i-1}]}.$$

For long strings of symbols, the average number of significant bits needed to represent r comes close to $H(X)$ bits per symbol.

Although the principle is fairly easy to describe, the implementation of arithmetic coding is not trivial. The floating-point registers of commonly used microprocessors cannot contain r with enough precision to encode a large stream of data. Suitable representations of r must be used as the encoding progresses. For more details, see, e.g., [106].

3.2 Joint and conditional entropies

I have defined the entropy of a single variable X together with its operational interpretation in source coding. Let me now describe some other useful entropic quantities.

First consider the case of two random variables X and Y. Together, they define a joint random variable XY with probability distribution $P_{XY}(x,y)$. These two random variables may be correlated, that is, in general the joint distribution is not the product of the marginal distributions; $P_{XY}(x,y) \neq P_X(x)P_Y(y)$. Then the entropy of both variables is defined as

$$H(X,Y) = -\sum_{x,y} P_{XY}(x,y) \log P_{XY}(x,y)$$

and satisfies $H(X,Y) \leq H(X) + H(Y)$ with equality if and only if X and Y are independent. Following a source coding interpretation, we can better compress the two variables jointly than each variable separately, since intuitively in the latter case their correlations would be coded twice.

The entropy can, of course, be easily extended to any number of variables. In general $H(X_1, \ldots, X_n)$ is the entropy of the random variable with the joint probability distribution $P_{X_1, \ldots, X_n}(x_1, \ldots, x_n)$.

The *conditional entropy*, denoted $H(X|Y)$, is defined as

$$H(X|Y) = H(X,Y) - H(Y).$$

This quantity characterizes the compressibility of the variable X if Y is known to both the encoder and the decoder. For a fixed value of Y, say $Y = y$, we denote as $X|Y = y$ the random variable on set $\mathcal{X}$ with probability distribution $P_{X|Y=y}(x) = P_{XY}(x,y)/P_Y(y)$. Assume that the encoder and the decoder have agreed on a different encoding scheme for each possible value of y, so as to take advantage of the differences in distributions that Y induces. For an optimal encoder, the encoding rate of the code associated with $Y = y$ is equal to $H(X|Y = y)$. On the other hand, the global rate of the encoder depends on the occurrences of Y. Hence, the global rate is the weighted average of $H(X|Y = y)$ with weight $P_Y(y)$. An elegant property of the conditional entropy is that it is precisely the global rate we are looking for:

$$H(X|Y) = \sum_{y \in \mathcal{Y}} P_Y(y) H(X|Y = y).$$

Note that the conditional entropy satisfies $H(X|Y) \leq H(X)$ with equality if and only if X and Y are independent. Intuitively, this may be understood as we cannot compress X worse with access to Y than without it.

3.3 Channel coding

Along with source coding, *channel coding* is the most important question addressed by information theory. It consists in finding the most efficient way to transmit information over a potentially noisy channel. The criterion to optimize is the transmission rate (i.e., we want to transmit as many symbols as possible per channel use) constrained by the transmission reliability (i.e., the probability that the message arrives inconsistently at the output of the channel must be vanishingly small).

A channel is characterized by an input alphabet $\mathcal{X}$, the symbols that the sender can transmit, and an output alphabet $\mathcal{Y}$, the symbols that the receiver gets. For simplicity, we will only discuss memoryless channels, that is, channels for which the behavior of one channel use does not depend on previous channel uses. In this setting, we can model the channel by a probability transition matrix $p(y|x)$, which expresses the probability of observing the output symbol y given that the input symbol is x. This transition matrix accounts for all the events that can happen during transmission: errors, substitutions, losses, etc.

In channel coding, an important quantity is the *mutual information* between two random variables. The mutual information between X and Y is written as $I(X;Y)$ and is defined as

$$I(X;Y) = H(X) + H(Y) - H(X,Y) = I(Y;X).$$

It satisfies $I(X;Y) \geq 0$ with equality iff X and Y are independent.

In terms of random variables, suppose that Alice sends to Bob some symbols, modeled by X, through the channel. Due to the channel, some changes to the transmitted variables occur as modeled by $p(y|x)$, and the values received by Bob are modeled by the random variable Y. Alice wishes to transmit a certain number of bits reliably through this channel. Shannon showed that the fundamental upper bound on the rate of transmission is $I(X;Y)$ bits per channel use. Of course, Alice can adapt the input distribution $P_X(x)$ to the given channel, thereby maximizing the transmission rate. This maximization defines the *channel capacity*

$$C = \max_{P_X(x)} I(X;Y).$$

For binary input alphabets, there are two important examples of channels, namely the binary erasure channel and the binary symmetric channel.

The binary erasure channel simply models losses in the transmission, for instance, due to packet losses over a network connection. While the input of the binary erasure channel is $\mathcal{X} = \{0,1\}$, its output is ternary, $\mathcal{Y} =$

$\{0, 1, ?\}$ with ? denoting an erasure. Erasure happens with a probability e, independently of the input symbol. Besides erased symbols, the channel is otherwise noiseless and the transition probability matrix reads:

$$p(0|0) = p(1|1) = 1 - e,$$
$$p(?|0) = p(?|1) = e,$$
$$p(1|0) = p(0|1) = 0.$$

It is easy to verify that the capacity of the binary erasure channel is $C_{\text{BEC}} = 1 - e$ bits per channel use. The input distribution that achieves capacity is the uniform distribution.

For a binary symmetric channel, the input and output alphabets are both binary, $\mathcal{X} = \mathcal{Y} = \{0, 1\}$. This channel models transmission errors. No erasures occur, but bits are flipped with a probability e, independently of the input symbol. Hence, the transition probability matrix reads:

$$p(0|0) = p(1|1) = 1 - e,$$
$$p(1|0) = p(0|1) = e.$$

The capacity of the binary symmetric channel is $C_{\text{BSC}} = 1 - h(e)$ bits per channel use. Again, the input distribution that achieves capacity is the uniform distribution.

3.3.1 Error-correcting codes

Error-correcting codes are methods to encode information in such a way that they are made resistant against errors caused by the channel over which they are transmitted. The well-known constructions of such codes follow the model given below. We shall restrict ourselves to the most-studied class of error-correcting codes, that is, linear codes, as they allow for facilitated encoding and decoding procedures without sacrificing efficiency.

Let $\text{GF}(2)^l$ be the vector space containing all l-bit vectors. A *linear error-correcting code* C is an l'-dimensional subspace of $\text{GF}(2)^l$. The subspace is spanned by a basis $\{g_i\}_{i=1...l'}$, and the encoding of the l'-bit symbol x consists in calculating the vector whose coordinates are (x_i) in the basis $\{g_i\}$: $c = \sum_i x_i g_i$. Alternatively, the subspace of the code can be represented by a *parity-check* matrix H: a codeword $c \in \text{GF}(2)^l$ belongs to C iff $Hc = 0$. Of course, $Hg_i = 0$ for $i = 1 \ldots l'$.

For any codeword c, we call the *syndrome* the value Hc. The syndrome of a codeword in the code C is always zero. Imagine that the codeword is sent through a binary symmetric channel; the syndrome can then be non-zero if the codeword undergoes an error. If the codeword c encounters an error $c \to c' = c + \epsilon$, then the syndrome becomes $H\epsilon$.

The standard error correction procedure first looks for ϵ', the *co-set leader* of $H\epsilon$, that is, the minimal-weight error pattern ϵ' such that $H\epsilon' = H\epsilon$. The weight of a binary vector is simply the number of non-zero components. In the case of a binary symmetric channel with $e < 1/2$, the most probable error pattern is also the lowest-weighted one. Then, the correction is applied to the codeword, $c' \to c'' = c' + \epsilon'$. If the error pattern was correctly recognized, we get $\epsilon = \epsilon'$ and $c'' = c$.

A codeword c is said to be orthogonal to some codeword c' iff $c \cdot c' = \sum_j c_j c'_j = 0$ in GF(2). Along with a code C, we define the *orthogonal code* as the vector space containing codewords $c^\perp$ orthogonal to all codewords of C. Formally,

$$C^\perp = \{c^\perp \in \text{GF}(2)^l : c^\perp \cdot c = 0 \, \forall c \in C\}.$$

The parity check matrix H of C contains rows that make a basis of $C^\perp$. Alternatively, the parity matrix $H^\perp$ of $C^\perp$ contains a basis of C.

3.3.2 Markov chains

The three random variables $X \to Y \to Z$ are said to form a *Markov chain* (in that order) if the joint probability distribution can be written as

$$P_{XYZ}(x,y,z) = P_X(x) P_{Y|X}(y|x) P_{Z|Y}(z|y).$$

An example of a Markov chain occurs when we connect two channels. Imagine that we send the random variable X through a first channel, which outputs Y, and that we send Y through a second channel, whose output yields Z. It is clear from the definition that $X \to Y \to Z$ is a Markov chain.

The consequence of $X \to Y \to Z$ being a Markov chain on mutual information is that $I(X;Y) \geq I(X;Z)$ and $I(X;Y|Z) \leq I(X;Y)$, with $I(X;Y|Z) = H(X|Z) + H(Y|Z) - H(X,Y|Z)$. We will need these inequalities in the sequel.

3.4 Rényi entropies

The Rényi entropies [148] form a family of functions on the probability distributions, which generalize (and include) the Shannon entropy. Their operational interpretation is less direct than for the Shannon entropy. Yet, they are of great importance for secret-key distillation – see Chapter 6.

The *Rényi entropy* of order r, with $0 < r < \infty$ and $r \neq 1$, of X is defined as

$$H_r(X) = \frac{1}{1-r} \log \sum_x (P_X(x))^r.$$

For $r = 0, 1, \infty$, we conventionally define

$$H_0(X) = \log |\{x \in \mathcal{X} : P_X(x) > 0\}|,$$

the logarithm of the support size of X;

$$H_1(X) = H(X),$$

the regular Shannon entropy; and

$$H_\infty(X) = -\log \max_x P_X(x),$$

the negative logarithm of the largest symbol probability.

An important particular case is the order-2 Rényi entropy

$$H_2(X) = -\log \sum P_X^2(x),$$

which is in fact the negative logarithm of the collision probability. The collision probability $\sum P_X^2(x)$ is the probability that two independent realizations of the random variable X are equal. For a random variable U with uniform distribution (i.e., $P_U(u) = |\mathcal{U}|^{-1}$ for all $u \in \mathcal{U}$), the order-2 Rényi and Shannon entropies match: $H_2(U) = H(U) = \log |\mathcal{U}|$. For any other random variable, the order-2 Rényi entropy is smaller than the Shannon entropy, i.e., $H_2(X) < H(X)$ with X non-uniform.

These properties can, in fact, be generalized to the rest of the family. For a given variable X, the Rényi entropies are non-increasing with respect to the order: $H_r(X) \leq H_s(X)$ iff $r \geq s$. The uniform distribution yields the same quantity for all Rényi entropies: $H_r(U) = \log |\mathcal{U}|$ for all $0 \leq r \leq \infty$. For non-uniform random variables, the Rényi entropy is strictly decreasing with respect to the order: $r > s \Leftrightarrow H_r(X) < H_s(X)$ for X non-uniform.

The joint Rényi entropy of multiple random variables is calculated over their joint probability distribution. Like the Shannon entropy, they satisfy

$$H_r(X, Y) \leq H_r(X) + H_r(Y).$$

The conditional Rényi entropy can also be defined. For two random variables $X \in \mathcal{X}$, $Y \in \mathcal{Y}$ and for some $y \in \mathcal{Y}$, let $H_r(X|Y = y)$ be the order-r Rényi entropy of the random variable $X|Y = y$ with distribution $P_{X|Y=y}(x)$

over $\mathcal{X}$. Then, $H_r(X|Y)$ is defined as

$$H_r(X|Y) = \sum_{y \in \mathcal{Y}} P_Y(y) H_r(X|Y=y).$$

3.5 Continuous variables

So far, we have only considered discrete random variables. In this section, we treat the case of continuous random variables, which allow us to model quantities that can vary continuously. In particular, they will be necessary when describing the quantum distribution of continuous key elements.

3.5.1 Differential entropy

The *differential entropy* of a continuous random variable X is defined as

$$H(X) = -\int_{\mathcal{X}} \mathrm{d}x \, p_X(x) \log p_X(x).$$

Note that the differential entropy is defined up to the scaling of the random variable. For simplicity, let us assume that X takes real values, $\mathcal{X} \subseteq \mathbf{R}$. Then, for some real number $a \neq 0$, let aX be the random variable that takes the value ax whenever $X = x$. Then, $H(aX) = H(X) + \log|a|$. This is a fundamental difference between differential entropy and the Shannon entropy: the differential entropy is sensitive to an invertible transformation of the symbols, while the Shannon entropy is not. As a consequence, there is no guarantee for the differential entropy to be positive: $H(X) < 0$ can happen.

The conditional differential entropy is naturally defined as $H(X|Y) = H(X, Y) - H(Y)$ for continuous random variables X and Y. Following the same argument as above, the conditional differential entropy can be negative.

The mutual information follows the same semantic definition as for discrete variables,

$$I(X;Y) = H(X) + H(Y) - H(X,Y).$$

The mutual information still satisfies $I(X;Y) \geq 0$ even when X and Y are continuous variables. If the variables X and Y are scaled by the real numbers $a, b \neq 0$, respectively, the logarithms of the scaling factors cancel each other, so that $I(aX; bY) = I(X; Y)$. In fact, the mutual information is insensitive to bijective transformations of X and Y: $I(f(X); g(X)) = I(X; Y)$ for f and g bijective functions over $\mathcal{X}$ and $\mathcal{Y}$, respectively. As a consequence, the channel interpretation of the mutual information as in Section 3.3 is still valid for continuous variables.

3.5.2 Gaussian variables and Gaussian channels

An important example of a continuous random variable is the *Gaussian random variable*, which is a real-valued random variable with a Gaussian distribution. Let $X \sim N(\mu, \Sigma)$ be a Gaussian variable with mean μ and standard deviation Σ, i.e.,

$$p_X(x) = \frac{1}{\Sigma\sqrt{2\pi}} e^{-\frac{(x-\mu)^2}{2\Sigma^2}}.$$

The differential entropy of X is

$$H(X) = 2^{-1} \log(2\pi e \Sigma^2).$$

Let X be transmitted through a *Gaussian channel*, that is, a channel which adds a Gaussian noise $\epsilon \sim N(0, \sigma)$ of standard deviation σ on the signal, giving $Y = X + \epsilon$ as output. Conditional on X, the output Y is distributed as a Gaussian with standard deviation σ, so that the entropy of Y conditional on X becomes $H(Y|X) = 2^{-1}\log(2\pi e \sigma^2)$ bits. The distribution of Y is Gaussian with variance $\Sigma^2 + \sigma^2$ and thus $H(Y) = 2^{-1}\log(2\pi e(\Sigma^2 + \sigma^2))$ bits. Consequently, the mutual information between X and Y reads

$$I(X;Y) = H(Y) - H(Y|X) = \frac{1}{2}\log\left(1 + \frac{\Sigma^2}{\sigma^2}\right), \qquad (3.1)$$

where Σ^2/σ^2 is called the *signal-to-noise ratio* (snr).

Note that a Gaussian channel can transmit an arbitrarily high number of bits if the input distribution has a sufficiently high standard deviation Σ. In practice, the channel models a physical medium and in this context, the variance Σ^2 is proportional to the energy transmitted through the medium. We thus naturally impose a constraint on the variance of the input variable, i.e., $\Sigma \leq \Sigma_{\max}$.

In fact, the Gaussian distribution yields the best rate for a given variance. So the capacity of the Gaussian channel is indeed the expression of Eq. (3.1) with $\Sigma = \Sigma_{\max}$.

3.6 Perfect secrecy revisited

Now that we have become more familiar with entropies, we can rephrase some aspects of perfect secrecy that were explained in Section 2.1.1. We say that a cipher achieves perfect secrecy when $\Pr[P = p|C = c] = \Pr[P = p]$ for all $c \in \mathcal{C}, p \in \mathcal{P}$, with $P \in \mathcal{P}$ (or $C \in \mathcal{C}$) is the random variable modeling the plaintext (or ciphertext). Equivalently, perfect secrecy is achieved when the knowledge of an eavesdropper (Eve) on P does not increase when she sees the ciphertext. In terms of entropy, this translates to $H(P|C) = H(P)$ or, equivalently, to

$$\text{perfect secrecy} \Leftrightarrow I(P; C) = 0.$$

If we see the encryption algorithm as an abstract transmission channel, no bit can be transmitted through this channel; there is thus no information given by C on P.

In 1949, Shannon proved [161] that perfect secrecy requires at least as many key bits as the message,

$$\text{perfect secrecy} \Rightarrow H(K) \geq H(P),$$

with K the random variable modeling the key. In the context of encryption with the one-time pad, this confirms that we need at least as many key bits as the message. Furthermore, we may not reuse previous keys and the successive outcomes of K must be independent of each other.

Let us now look at the case where Eve has a small, yet non-zero, amount of information on the key. We look at the key, plaintext and ciphertext as blocks of l bits, not bit per bit. We assume that the key is chosen completely randomly, so that $H(K) = l$. Let Z be the random variable modeling Eve's knowledge on the key, and let her knowledge be quantified as $I(K; Z) = \epsilon$. Then, we find:

$$H(P|C, Z) = H(P, C|Z) - H(C|Z) \tag{3.2}$$
$$= H(P, K|Z) - H(P \oplus K|Z) \tag{3.3}$$
$$\geq H(P|K, Z) + H(K|Z) - l \tag{3.4}$$
$$= H(P) - \epsilon. \tag{3.5}$$

The knowledge ϵ on the key must be considered as a weakening of the confidentiality of the plaintext, since its a-posteriori entropy can be decreased by the same amount. It is, therefore, essential to be able to control ϵ and to reduce it below an acceptable limit. Be aware, however, that requiring a small amount of Shannon information ϵ does not mean a small number of bits eavesdropped. It can also mean that most of the time, Eve gets no information whatsoever, but a fraction ϵ/l of the time, she gets all the bits.

Note that the inequality in Eq. (3.4) becomes an equality if $H(P) = l$, that is, if the plaintext is ideally compressed before encryption. In this case, $I(P; C, Z) = \epsilon$.

3.7 Conclusion

I have given the important concepts of information theory such as source coding and channel coding. I also defined important quantities such as the Shannon entropy, the mutual information, the Rényi entropy and the differential entropy. Finally, perfect secrecy was covered again in the light of the new information-theoretic concepts.

For more information, a good introduction to information theory can be found in the book by Cover and Thomas [46].

4
Quantum information theory

In this chapter, I will introduce the concepts of quantum information that are necessary to the understanding of quantum cryptography. I first review the main principles of quantum mechanics. I then introduce concepts that somehow translate information theory into the quantum realm and discuss their unique features. Finally, I conclude this chapter with some elements of quantum optics.

4.1 Fundamental definitions in quantum mechanics

In classical mechanics, a system is described with physical quantities that can take certain values at a given moment in time; we say that is has a state. For instance, the state of an elevator comprises its position and speed at a given time. As dictated by common sense, if the elevator is at a given height, it cannot be simultaneously found at another location. With quantum mechanics, things are fundamentally different and elementary particles can behave against common sense. For instance, a quantum system can simultaneously be in different levels of energy. This is not because our knowledge of the state of the system is incomplete: the behavior of the quantum system is consistent with the fact that it is in simultaneous levels of energy.

It would go beyond the scope of this book to describe quantum mechanics in detail. Instead, I propose a synthetic introduction suitable for the understanding of quantum cryptography. For a more detailed introduction, please refer to the books listed at the end of this chapter.

4.1.1 Quantum states

In (non-relativistic) quantum mechanics, a physical system is described as a complex Hilbert (vector) space $\mathcal{H}$. The state of such a physical system is any unit-sized vector, denoted $|\psi\rangle$ in the standard Dirac notation.

Assuming a finite or countable number of dimensions of $\mathcal{H}$, the Hilbert space is described as spanned by some orthonormal basis $\{|a\rangle\}$. We can then write $|\psi\rangle$ as a complex linear combination of the basis vectors:

$$|\psi\rangle = \sum_a c_a |a\rangle, \text{ with } c_a \in \mathbf{C} \text{ and } \sum_a |c_a|^2 = 1.$$

For an uncountable number of dimensions, the idea remains the same with sums replaced by integrals.

The decomposition coefficients can be expressed as $c_a = \langle a|\psi\rangle$, the inner product between $|a\rangle$ and $|\psi\rangle$. More generally, the inner product between two states $|\psi_1\rangle$ and $|\psi_2\rangle$ is denoted as $\langle\psi_1|\psi_2\rangle$. Using the basis vectors $\{|a\rangle\}$ and the decomposition $|\psi_i\rangle = \sum_a c_{i,a}|a\rangle$, $i = 1, 2$, the inner product of two vectors reads

$$\langle\psi_1|\psi_2\rangle = \sum_a \langle\psi_1|a\rangle \times \langle a|\psi_2\rangle = \sum_a c^*_{1,a} c_{2,a}.$$

Note that the notation $\langle\psi|$ represents a linear application from $\mathcal{H}$ to $\mathbf{C}$. The inner product verifies the property that $\langle\psi_2|\psi_1\rangle = \langle\psi_1|\psi_2\rangle^*$.

A linear combination such as $|\psi\rangle = \sum_a c_a |a\rangle$ is called a *superposition*. If the states $|a\rangle$ each have a classical equivalent (e.g., the state $|a\rangle$ means that the system has energy a), the superposition of such states cannot be translated into classical terms (e.g., the state $2^{-1/2}|a\rangle + 2^{-1/2}|b\rangle$ means that the system is in both energy levels a and b).

4.1.2 Measurements

In classical mechanics, measuring the state of a system is often implicit. For instance, we can look at a ball rolling on the floor and note its position and speed at each moment of time. Our common sense says that looking at the ball does not modify its trajectory. Hence, the physics of the observer can be completely decoupled from the physics of the system under investigation.

In contrast, measurements are an integral part of the quantum mechanics postulates. This is because a measurement has an incidence on the quantum system. We cannot simply look at an elementary particle. Looking at the particle would mean that photons are reflected by the particle, hence creating an interaction between the photons and the system. Surprisingly,

4.1 Fundamental definitions in quantum mechanics

the measurement perturbs the system, even if our apparatus is perfect and no matter how gently we interact with the system.

In quantum mechanics, physical quantities are associated with linear operators called *observables*. A linear operator is a linear application from $\mathcal{H}$ to $\mathcal{H}$. The application of an operator $\mathbf{O}$ to a state $|\psi\rangle$ is denoted as $\mathbf{O}|\psi\rangle$. The inner product between a state $|\psi_1\rangle$ and another state $|\psi_2\rangle$ transformed by some operator $\mathbf{O}$ is denoted as $\langle\psi_1|\mathbf{O}|\psi_2\rangle$ and is called a *matrix element*.

Actually, observables are Hermitian operators: an operator $\mathbf{O}$ is Hermitian iff $\langle\psi|\mathbf{O}|\phi\rangle = \langle\phi|\mathbf{O}|\psi\rangle^*$ for all $|\psi\rangle$ and $|\phi\rangle$. The eigenvalues of Hermitian operators are real.

For simplicity, I will describe only discrete orthogonal measurements – more general measurements can be found in standard textbooks. Let the observable $\mathbf{O}$ have a finite or countable set of eigenvectors, making an orthonormal basis $\{|b\rangle\}$ of $\mathcal{H}$. Each eigenvector $|b\rangle$ has the associated real eigenvalue λ_b. It is a postulate of quantum mechanics that the measurement of $\mathbf{O}$ on a state $|\psi\rangle$ gives the result λ with probability

$$p_\lambda = \sum_{b:\lambda_b=\lambda} |\langle b|\psi\rangle|^2.$$

The average result $\sum p_\lambda \lambda$ is denoted $\langle\mathbf{O}\rangle$. After the measurement, the state undergoes a projection onto the eigenspace of λ:

$$|\psi'\rangle = \frac{1}{p_\lambda} \sum_{b:\lambda_b=\lambda} |b\rangle\langle b|\psi\rangle.$$

Hence, in general, the measurement modifies the state.

With two operators $\mathbf{O}_1$ and $\mathbf{O}_2$, we define their *commutator* as

$$[\mathbf{O}_1, \mathbf{O}_2] = \mathbf{O}_1\mathbf{O}_2 - \mathbf{O}_2\mathbf{O}_1.$$

We say that $\mathbf{O}_1$ and $\mathbf{O}_2$ *commute* when $[\mathbf{O}_1, \mathbf{O}_2] = 0$. Non-commuting observables cannot yield measurement results with perfect precision simultaneously. Let $\Delta\mathbf{O}$ be the standard deviation of the measurement results of $\mathbf{O}$ on identically-prepared states $|\psi\rangle$. The *Heisenberg uncertainty principle* reads

$$\Delta\mathbf{O}_1\Delta\mathbf{O}_2 \geq |\langle[\mathbf{O}_1, \mathbf{O}_2]\rangle|/2, \tag{4.1}$$

meaning that both standard deviations cannot be simultaneously arbitrarily small for non-commuting observables.

4.1.3 Evolution

Quantum systems can evolve with time, whether this evolution is caused by the environment or by some apparatus. I will not describe the nature of this evolution, as it strongly depends on the physics of the quantum system and of the apparatus or of the environment it interacts with. Instead, I will only describe the macroscopic properties of quantum state evolution.

If the quantum state at time t is $|\psi\rangle$, the quantum state at $t + \Delta t$ is $\mathbf{U}|\psi\rangle$ for some unitary operator $\mathbf{U}$. A *unitary operator* $\mathbf{U}$ is an invertible linear operator that preserves the norm of the vectors. Formally, it verifies $\mathbf{U}^\dagger \mathbf{U} = \mathbf{U}\mathbf{U}^\dagger = \mathbf{I}$, with $\mathbf{I}$ the identity operator.

Unitarity implies that $\mathbf{U}|\psi\rangle$ is still a valid quantum states, i.e., it has unit size if $|\psi\rangle$ has unit size. Also, the evolution with unitary operators is reversible: it is always possible to come back to the previous state by undoing the unitary operator.

Note that the reversibility of evolution may seem to contradict the fact that measurements are irreversible. If, however, one considers the joint quantum system made of the system under measurement *and* of the apparatus itself, then the measurement can be described as a unitary operator. From the point of view of the quantum system being measured only, measurement is irreversible.

4.2 Qubits and qubit pairs

While the bit models the smallest classical information unit, the qubit models the smallest quantum information unit. A *qubit* is a quantum system that lies in a two-dimensional Hilbert space, $\dim \mathcal{H} = 2$. The basis of $\mathcal{H}$ is denoted as $\{|0\rangle, |1\rangle\}$ and is sometimes called the *computational basis*. A qubit state is thus described as $c_0|0\rangle + c_1|1\rangle$ with $|c_0|^2 + |c_1|^2 = 1$. In a sense, a classical bit is a qubit restricted to the basis states.

4.2.1 Pauli operators

For qubits, there are three fundamental operators, called Pauli operators or Pauli matrices:

$$\mathbf{X} = \begin{pmatrix} 0 & 1 \\ 1 & 0 \end{pmatrix}, \mathbf{Y} = \begin{pmatrix} 0 & -i \\ i & 0 \end{pmatrix} = i\mathbf{XZ}, \mathbf{Z} = \begin{pmatrix} 1 & 0 \\ 0 & -1 \end{pmatrix}.$$

The basis vectors $\{|0\rangle, |1\rangle\}$ are eigenvectors of $\mathbf{Z}$. Following the postulates described in Section 4.1.2, measurement of the state $|\psi\rangle = c_0|0\rangle + c_1|1\rangle$ with the observable $\mathbf{Z}$ yields the result $|0\rangle$ with probability $|c_0|^2$ and $|1\rangle$ with probability $|c_1|^2$.

The eigenstates of $\mathbf{X}$ are $|+\rangle$ with eigenvalue 1 and $|-\rangle$ with eigenvalue -1, where

$$|+\rangle = 2^{-1/2}(|0\rangle + |1\rangle) \text{ and } |-\rangle = 2^{-1/2}(|0\rangle - |1\rangle).$$

Note that $|\psi\rangle = c_0|0\rangle + c_1|1\rangle = c_+|+\rangle + c_-|-\rangle$ with $c_+ = 2^{-1/2}(c_0 + c_1)$ and $c_- = 2^{-1/2}(c_0 - c_1)$. Measurement of $|\psi\rangle$ with $\mathbf{X}$ yields $|+\rangle$ with probability $|c_+|^2$ and $|-\rangle$ with probability $|c_-|^2$.

In BB84, Alice sends states from the set $\{|0\rangle, |1\rangle, |+\rangle, |-\rangle\}$ and Bob measures either $\mathbf{X}$ or $\mathbf{Z}$ – see Section 1.1. From the derivations above, it is now clear that measuring $\mathbf{X}$ (or $\mathbf{Z}$) when Alice sends $|0\rangle$ or $|1\rangle$ (or $|+\rangle$ or $|-\rangle$) gives random results.

For completeness, note that the eigenstates of $\mathbf{Y}$ are

$$|0_y\rangle = 2^{-1/2}(|0\rangle + i|1\rangle) \text{ and } |1_y\rangle = 2^{-1/2}(|0\rangle - i|1\rangle).$$

The operator $\mathbf{X}$ is usually called the *bit-flip operator*, as the state $c_0|0\rangle + c_1|1\rangle$ becomes $c_0|1\rangle + c_1|0\rangle$. The operator $\mathbf{Z}$, on the other hand, is usually called the *phase-flip operator*, as the state $c_0|0\rangle + c_1|1\rangle$ becomes $c_0|0\rangle - c_1|1\rangle$.

4.2.2 Multiple qubit systems

Let us consider a system with n joint qubits. The basis of the system comprises 2^n vectors: $\{|00\ldots0\rangle, |00\ldots1\rangle, \ldots, |11\ldots1\rangle\}$. To describe the value of a classical n-bit vector, we obviously need n bits; to describe the state of n qubits, we need 2^n complex numbers. Unitary operators on n qubits hence influence 2^n complex numbers simultaneously – this is the kind of parallelism that quantum computers exploit.

As an important special case, let us consider a system of two qubits, that is, $\dim \mathcal{H} = 2^2 = 4$. In such a system, we can use the 2-qubit computational basis $\{|00\rangle, |01\rangle, |10\rangle, |11\rangle\}$. Another useful basis is the basis of *maximally entangled states* or the *Bell basis*:

$$|\phi^+\rangle = 2^{-1/2}(|00\rangle + |11\rangle), \qquad |\phi^-\rangle = 2^{-1/2}(|00\rangle - |11\rangle),$$
$$|\psi^+\rangle = 2^{-1/2}(|01\rangle + |10\rangle), \qquad |\psi^-\rangle = 2^{-1/2}(|01\rangle - |10\rangle).$$

This basis is useful for many reasons that will appear clearly later. Let us simply note here that applying a bit-flip on the right qubit of $|\phi^+\rangle$ yields $|\psi^+\rangle = (\mathbf{I} \otimes \mathbf{X})|\phi^+\rangle$, a phase-flip yields $|\phi^-\rangle = (\mathbf{I} \otimes \mathbf{Z})|\phi^+\rangle$ and both flips yield $|\psi^-\rangle = (\mathbf{I} \otimes \mathbf{XZ})|\phi^+\rangle$.

When measuring either part of a maximally entangled state, the results yield correlations that have no classical equivalence. They are sometimes also called Einstein–Podolsky–Rosen (EPR) correlations, from the famous paper by these three physicists [56].

4.3 Density matrices and quantum systems

Classical uncertainties come on top of the quantum description of a state. In BB84, for instance, Alice randomly emits one of the states $|0\rangle$, $|1\rangle$, $|+\rangle$, $|-\rangle$. From the point of view of an observer, the system at the output of her station cannot be described by a quantum state as we have seen up to now.

Both quantum superpositions and classical uncertainties can be unified in a single terminology with objects called *density matrices*. A density matrix is a positive linear operator that represents a quantum state. Note that the density operator verifies $0 \leq \langle\psi|\rho|\psi\rangle \leq 1$ for all $|\psi\rangle$. In this new language, a vector $|\psi\rangle$ is translated as the density operator $\rho = |\psi\rangle\langle\psi|$ called a projector.

A classical random variable, yielding state $|\psi_i\rangle$ with probability p_i, gives the density matrix

$$\rho = \sum_i p_i |\psi_i\rangle\langle\psi_i|.$$

We say that a quantum state ρ is *pure* whenever it can be written as a projector $\rho = |\psi\rangle\langle\psi|$. Otherwise, it is *mixed*.

For instance, the states $|0\rangle\langle 0|$, $|1\rangle\langle 1|$, $|+\rangle\langle +|$ and $|-\rangle\langle -|$ are pure. What Alice sends in BB84 is the mixed state

$$\rho = \frac{1}{4}(|0\rangle\langle 0| + |1\rangle\langle 1| + |+\rangle\langle +| + |-\rangle\langle -|) = \mathbf{I}/2.$$

The *trace* of an operator $\operatorname{Tr}\mathbf{O}$ is defined as

$$\operatorname{Tr}\mathbf{O} = \sum_a \langle a|\mathbf{O}|a\rangle$$

for some orthonormal basis $\{|a\rangle\}$. Note that the trace is independent of the chosen basis. For a density matrix, we always have $\operatorname{Tr}\rho = 1$.

With the density matrix notation, the average result of a measurement $\mathbf{O}$ on state ρ reads $\langle \mathbf{O}\rangle = \operatorname{Tr}(\mathbf{O}\rho)$. This average takes into account both the quantum superpositions and the classical uncertainties.

Let us now consider two physical systems a and b represented by the Hilbert spaces $\mathcal{H}_\mathsf{a}$ and $\mathcal{H}_\mathsf{b}$ respectively. The joint system is represented by the Hilbert space $\mathcal{H} = \mathcal{H}_\mathsf{a} \otimes \mathcal{H}_\mathsf{b}$ with basis $\{|a\rangle_\mathsf{a}|b\rangle_\mathsf{b}\ \forall a,b\}$. For a state $\rho = \sum \rho_{aa'bb'} |a\rangle_\mathsf{a}\langle a'| \otimes |b\rangle_\mathsf{b}\langle b'|$, the *partial trace* over the system b is defined

as
$$\rho_\mathsf{a} = \mathrm{Tr}_\mathsf{b}\rho = \sum_{aa'b} \rho_{aa'bb}|a\rangle_\mathsf{a}\langle a'|.$$

In a system **ab**, the partial trace over **b** models the system as an observer would perceive it when only having access to the part **a**. The system **ab** being in a pure state $\rho = |\psi\rangle_\mathsf{ab}\langle\psi|$ does not imply that the partial system **a** is also in a pure state (i.e., $\rho_\mathsf{a} \neq |\psi\rangle_\mathsf{a}\langle\psi|$ in general). On the other hand, given a state ρ_a, we can always find a larger system **ab** such that $\rho_\mathsf{a} = \mathrm{Tr}_\mathsf{b}|\psi\rangle_\mathsf{ab}\langle\psi|$ for some appropriately chosen pure state $|\psi\rangle_\mathsf{ab}\langle\psi|$.

A state ρ in the system **ab** that can be written as $\rho = \rho_\mathsf{a} \otimes \rho_\mathsf{b}$ is said to be *separable*. A non-separable state $\rho \neq \rho_\mathsf{a} \otimes \rho_\mathsf{b}$ is *entangled*.

To measure some form of distance (or closeness) between two states, one defines the *fidelity*. The fidelity between a pure state $|\psi\rangle\langle\psi|$ and a mixed state ρ is defined as
$$F(|\psi\rangle\langle\psi|, \rho) = \langle\psi|\rho|\psi\rangle.$$

The fidelity verifies the inequality $0 \leq F(|\psi\rangle\langle\psi|, \rho) \leq 1$ and the property that $F(|\psi\rangle\langle\psi|, \rho) = 1$ iff $|\psi\rangle\langle\psi| = \rho$.

4.4 Entropies and coding

Whereas the Shannon entropy measures the uncertainty of a classical random variable, the von Neumann entropy measures some form of uncertainty in a quantum state.

The *von Neumann entropy* of state ρ is defined as
$$H(\rho) = -\mathrm{Tr}(\rho \log \rho) = -\sum_i \lambda_i \log \lambda_i,$$

with $\{\lambda_i\}$ the eigenvalues of ρ and where the logarithm is in base 2. The von Neumann entropy of a pure state is zero, $H(|\psi\rangle\langle\psi|) = 0$. It is always positive and is upper-bounded by the logarithm of the number of dimensions of the Hilbert space, $0 \leq H(\rho) \leq \log \dim \mathcal{H}$.

In the classical world, one can compress a classical variable X in a number of bits as close as desired to $H(X)$ bits per realization. By gathering d instances of the variables, one can describe them using approximately $dH(X)$ bits.

In the quantum world, *Schumacher compression* plays an analogue role [156]. If a large number d of identical quantum states $\rho^{\otimes d}$ are gathered, their support approximately fits in a $2^{dH(\rho)}$-dimensional Hilbert space. Even though these d states are described in a Hilbert space of dimension $(\dim \mathcal{H})^d$, there exists an encoding **E** that maps $\rho^{\otimes d}$ into a smaller Hilbert space while retaining most information. The Hilbert space of the encoded state is made of dR qubits, for some real number $R \leq \log \dim \mathcal{H}$. The encoded state can then be decoded using $\mathbf{E}^{-1}$, that is, expanded back into the initial Hilbert space. More precisely, the fidelity $F(\mathbf{E}^{-1}\mathbf{E}\rho^{\otimes d}, \rho^{\otimes d})$ between $\rho^{\otimes d}$ and the state encoded and then decoded can be as close to 1 as desired if $R > H(\rho)$ and if d is sufficiently large.

4.5 Particularity of quantum information

Let us review some properties of quantum information theory that have no equivalence in classical information theory. The first interesting property of quantum mechanics is the *impossibility to distinguish non-orthogonal states perfectly*.

First, let us consider the case of orthogonal states. We are given the description of two orthogonal states $|\psi_1\rangle$ and $|\psi_2\rangle$, with $\langle\psi_1|\psi_2\rangle = 0$, and an unknown state $|\psi\rangle \in \{|\psi_1\rangle, |\psi_2\rangle\}$. We wish to determine whether $|\psi\rangle = |\psi_1\rangle$ or $|\psi\rangle = |\psi_2\rangle$. For this purpose, we can construct an observable **O** such that $|\psi_1\rangle$ and $|\psi_2\rangle$ are eigenvectors with distinct eigenvalues λ_1 and λ_2, respectively. The measurement of **O** will thus give λ_i, $i \in \{1, 2\}$, as a result and we can infer that $|\psi\rangle = |\psi_i\rangle$. Since $|\langle\psi_1|\psi_2\rangle|^2 = 0$, the probability of a mismatch is zero. E.g., since $\langle 0|1\rangle = 0$, it is always possible to distinguish between $|0\rangle$ and $|1\rangle$; the same applies to $|+\rangle$ and $|-\rangle$.

Then, we consider the case of non-orthogonal states. Again, we are given the description of two states $|\psi_1\rangle$ and $|\psi_2\rangle$ such that $\langle\psi_1|\psi_2\rangle \neq 0$. Assume that we construct an observable that is able to distinguish between $|\psi_1\rangle$ and $|\psi_1^\perp\rangle$ with $\langle\psi_1|\psi_1^\perp\rangle = 0$, using the construction above. Now the probability of a mismatch is non-zero since $|\langle\psi_1|\psi_2\rangle|^2 \neq 0$. When the measurement outcome yields λ_1, there is no way to be sure that $|\psi\rangle = |\psi_1\rangle$ since $|\psi\rangle = |\psi_2\rangle$ is also possible. More general measurements can be investigated, but the result is always ambiguous in distinguishing between $|\psi_1\rangle$ and $|\psi_2\rangle$. E.g., since $\langle+|1\rangle \neq 0$, it is not possible to distinguish between $|+\rangle$ and $|1\rangle$.

In general, *an unknown state is disturbed after a measurement*. We must be reminded that after a measurement the state $|\psi\rangle$ is projected onto the eigenspace associated with λ, the result of the measurement. In particular, suppose that we are again given the state $|\psi\rangle = |\psi_2\rangle$. If we make a measurement that distinguishes between $|\psi_1\rangle$ and some orthogonal state $|\psi_1^\perp\rangle$, the probability that the result is λ_1 is non-zero. If this occurs, the unknown state $|\psi\rangle = |\psi_2\rangle$ is transformed into $|\psi'\rangle = |\psi_1\rangle$ after the measurement, and any subsequent measurement yields results in accordance with the new state

$|\psi'\rangle$. We thus cannot make several measurements on a single unknown state in the hope of contradicting the impossibility of perfectly distinguishing non-orthogonal states.

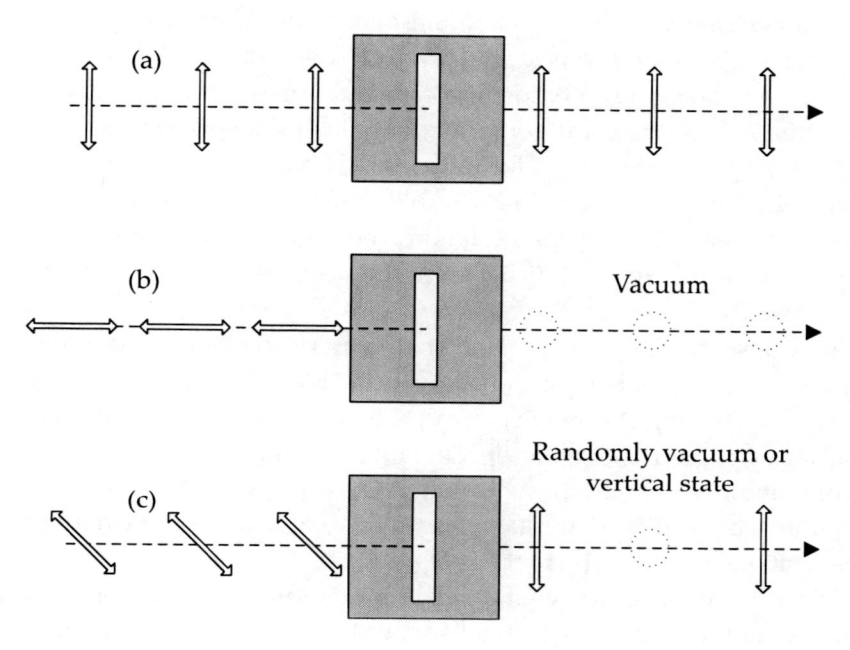

Fig. 4.1. A single photon impinging on a filter that only allows vertically polarized light to pass. (a) Vertically polarized photons pass through the filter without being absorbed. (b) Horizontally polarized photons are all absorbed. (c) Diagonally polarized photons are absorbed or transmitted at random. An observer placed after the filter cannot, therefore, determine in a deterministic manner the state of the photon before the filter, in contrast with the vertical–horizontal case. Furthermore, the photon is no longer diagonally polarized after the filter.

The impossibility of perfectly distinguishing non-orthogonal states and their disturbance after a measurement is illustrated for polarized light in Fig. 4.1.

This impossibility result is applicable only if a single instance of the unknown state $|\psi\rangle$ is given. If many instances of the same state $|\psi\rangle$ are given, it is always possible to find an observable yielding different result statistics for the two states $|\psi_1\rangle$ and $|\psi_2\rangle$. It then becomes a matter of accumulating enough data to be able to infer the identity of the unknown state statistically; this technique is called *quantum tomography*.

At this point, it may be tempting to say that we can take the unknown state $|\psi\rangle$ and make several copies of it. This way, we can circumvent the problem of distinguishing non-orthogonal states. However, making two or more perfect copies of an unknown state $|\psi\rangle$ is impossible; this is the *no-cloning theorem*.

The no-cloning theorem says that there does not exist any quantum operation – a *cloning machine* – that would make a perfect copy of an unknown state. More precisely, given a set $\{|\psi_i\rangle\}$ of states in $\mathcal{H}$ to clone and if any two states in the set are non-orthogonal, then there does not exist $\mathbf{U}$, acting within $\mathcal{H}^{\otimes 2} \otimes \mathcal{H}_{\mathsf{aux}}$ for some auxiliary Hilbert space $\mathcal{H}_{\mathsf{aux}}$, such that $\mathbf{U}|\psi_i\rangle|0\rangle|0\rangle_{\mathsf{aux}} = |\psi_i\rangle|\psi_i\rangle|i\rangle_{\mathsf{aux}}$ for all $|\psi_i\rangle$.

Although perfect cloning is impossible, approximate cloning is indeed allowed by the laws of quantum mechanics. For instance, it is possible to clone a qubit with a fidelity of 5/6 between the original state and either one of the two copies.

It may also be intuitively clear that a perfect cloning machine would contradict the Heisenberg uncertainty principle. For two non-commuting observables $\mathbf{O}_1$ and $\mathbf{O}_2$, making two perfect copies of $|\psi\rangle$ would allow one to make two measurements; one observable per copy. Each measurement on the copy number i would be such that $\Delta\mathbf{O}_i \to 0$ and $\Delta\mathbf{O}_{3-i} \to \infty$, $i = 1, 2$. Combining the results on both copies would give $\Delta\mathbf{O}_i \to 0$ for both $i = 1, 2$, hence contradicting Eq. (4.1).

To sum up, information contained in a quantum state in general cannot be copied and any attempt to read it can disturb the quantum state.

4.6 Quantum optics

Classically, a monochromatic plane wave can be written as

$$\bar{E}(\bar{r}, t) = -\frac{\mathcal{E}}{\sqrt{N_0}} \bar{\epsilon}(X \sin(\bar{k}\bar{r} - \omega t) + P \cos(\bar{k}\bar{r} - \omega t)), \qquad (4.2)$$

where $\bar{E}(\bar{r}, t)$ denotes the electric field at point in space $\bar{r}$ and at time t, $\mathcal{E}/\sqrt{N_0}$ is a constant, $\bar{\epsilon}$ is the polarization vector, $\bar{k}$ is the propagation direction, ω is the frequency, and X and P are called the *quadrature amplitudes*.

Such a source of light is rather a mathematical tool and does not exist by itself. In particular, it would contradict the laws of quantum mechanics. To remedy this, the quadrature amplitudes X and P must be replaced by the operators $\mathbf{x}$ and $\mathbf{p}$ in the Hilbert space $L^2(\mathbf{R})$.

The operators $\mathbf{x}$ and $\mathbf{p}$ are formally equivalent to the position and momentum operators in quantum mechanics. They verify the commutation relationship $[\mathbf{x}, \mathbf{p}] = 2iN_0$, where N_0 is the variance of the vacuum fluctuations – N_0 is also sometimes called the photon noise or the shot noise. We denote the eigenstates of $\mathbf{x}$ as $|x\rangle$ with $\mathbf{x}|x\rangle = x|x\rangle$ and $x \in \mathbf{R}$. The observable **x** and **p** can be measured by a technique called *homodyne detection*, see also Section 11.4.

The eigenstates $|x\rangle$ of **x** can be used to decompose any state in quantum optics:

$$|\psi\rangle = \int_{x \in \mathbf{R}} \mathrm{d}x f(x)|x\rangle,$$

with $\int \mathrm{d}x |f(x)|^2 = 1$ and $f(x) = \langle x|\psi\rangle$.

We can also define the operators

$$\mathbf{a} = \frac{\mathbf{x} + i\mathbf{p}}{2\sqrt{N_0}}, \quad \mathbf{a}^\dagger = \frac{\mathbf{x} - i\mathbf{p}}{2\sqrt{N_0}},$$

which are formally equivalent to the annihilation and creation operators of the harmonic oscillator. In the context of quantum optics, they annihilate and create photons. Finally, the *photon number operator* is $\mathbf{n} = \mathbf{a}^\dagger \mathbf{a}$.

4.6.1 Important families of states

Let us review some important families of states that we will use frequently in the sequel: the vacuum state, the photon number states, the coherent states and the squeezed states.

The *vacuum state* is denoted $|0\rangle$. It is an eigenstate of the annihilation operator with eigenvalue 0, namely $\mathbf{a}|0\rangle = 0$. It represents the absence of photons. The quadrature amplitudes have a zero average: $\langle \mathbf{x} \rangle = \langle \mathbf{p} \rangle = 0$. Surprisingly, the fluctuations are non-zero, but are by definition equal to N_0: $\langle \mathbf{x}^2 \rangle = \langle \mathbf{p}^2 \rangle = N_0 \neq 0$. The vacuum state is also a special case of photon number state and of coherent state (see below).

The *photon number states*, also called *Fock states*, are eigenstates of the photon number operator: $\mathbf{n}|n\rangle = n|n\rangle$, with $n \in \mathbf{N}$ a non-negative integer. The state $|n\rangle$ represents n photons. The photon number states form an orthonormal basis of the underlying Hilbert space, i.e., the Fock basis.

A *coherent state* is a state that has the same fluctuations of quadrature amplitudes as the vacuum state but which possibly has non-zero average quadrature amplitudes. For a complex number α, the coherent state $|\alpha\rangle$ satisfies $\langle \mathbf{x} \rangle = \mathrm{Re}\,\alpha$, $\langle \mathbf{p} \rangle = \mathrm{Im}\,\alpha$ and $\langle \mathbf{x}^2 \rangle = \langle \mathbf{p}^2 \rangle = N_0$. Note that some authors use another convention for α, where they apply a scaling of $1/2\sqrt{N_0}$ on α. The values $\langle \mathbf{x} \rangle$ and $\langle \mathbf{p} \rangle$ are called the displacement.

A coherent state is said to be Gaussian since in the basis of the eigenstates of **x**, it has a Gaussian shape:

$$\langle x|\alpha\rangle = (2\pi N_0)^{-1/4} e^{-\frac{(x-\langle \mathbf{x}\rangle)^2 + i\langle \mathbf{x}\rangle\langle \mathbf{p}\rangle}{4N_0} + \frac{i\langle \mathbf{p}\rangle x}{2N_0}}.$$

The coherent states are not orthogonal to themselves: $\langle \alpha|\beta\rangle = e^{-|\alpha-\beta|^2/4N_0}$.

A *squeezed state* is a Gaussian state with unequal fluctuations on either quadrature amplitude. For a parameter $s > 0$, called squeezing, the squeezed state $|\alpha, s\rangle$ satisfies $\langle \mathbf{x}\rangle = \operatorname{Re}\alpha$, $\langle \mathbf{p}\rangle = \operatorname{Im}\alpha$, $\langle \mathbf{x}^2\rangle = sN_0$ and $\langle \mathbf{p}^2\rangle = N_0/s$.

In the basis of the eigenstates of **x**, a squeezed state also has a Gaussian shape:

$$\langle x|\alpha, s\rangle = (2\pi N_0 s)^{-1/4} e^{-\frac{s^{-1}(x-\langle \mathbf{x}\rangle)^2 + i\langle \mathbf{x}\rangle\langle \mathbf{p}\rangle}{4N_0} + \frac{i\langle \mathbf{p}\rangle x}{2N_0}}.$$

Measuring the observable **x** (or **p**) of a coherent or squeezed state gives a Gaussian random variable with mean $\langle \mathbf{x}\rangle$ (or $\langle \mathbf{p}\rangle$) and variance $\langle \mathbf{x}^2\rangle$ (or $\langle \mathbf{p}^2\rangle$).

4.6.2 The Wigner function

The Wigner function is a nice tool for visualizing states in quantum optics. For a state ρ, the *Wigner function* $W(x, p)$ is defined as

$$W(x, p) = \frac{1}{4\pi N_0} \int \mathrm{d}q\, e^{\frac{ipq}{2N_0}} \langle x - \frac{q}{2}|\rho|x + \frac{q}{2}\rangle.$$

The Wigner function can be interpreted in the following way. If one measures **x**, the result follows a probability density function $p_X(x)$, which can be obtained by integrating the variable p of $W(x, p)$: $p_X(x) = \int \mathrm{d}p W(x, p)$. The symmetric result applies to the measurement of **p**: $p_P(p) = \int \mathrm{d}x W(x, p)$.

The Wigner function of a coherent state $|\alpha\rangle$ is a two-dimensional Gaussian distribution centered on $(\operatorname{Re}\alpha = \langle \mathbf{x}\rangle, \operatorname{Im}\alpha = \langle \mathbf{p}\rangle)$, with standard deviation $\sqrt{N_0}$ on both directions and with zero covariance. A coherent state is thus usually represented as a circle in the (x, p) plane, as depicted in Fig. 4.2.

The Wigner function of a squeezed state is similar to that of a coherent state, with different standard deviations however. One has to imagine that the x axis is scaled by a factor $s^{1/2}$ and the p axis by a factor $s^{-1/2}$. A squeezed state is thus usually represented as an ellipse in the (x, p) plane, as depicted in Fig. 4.2.

4.7 Conclusion

In this chapter, I introduced many important definitions and concepts in quantum physics that are necessary for the understanding of quantum crypto-

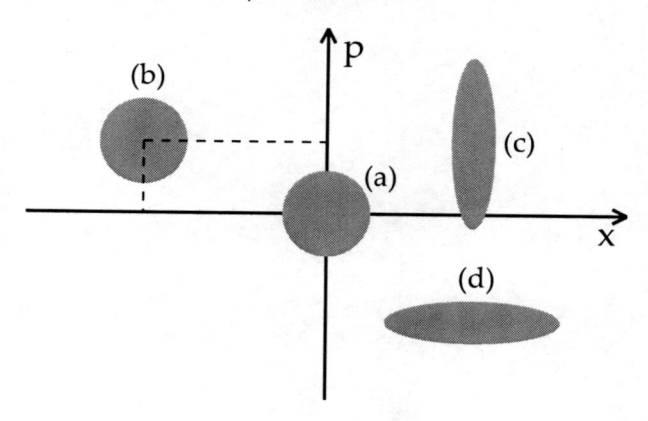

Fig. 4.2. Sketch of the Wigner function for four different states. (a) For the vacuum state $|0\rangle$, the Gaussian distribution is centered on the origin. (b) A displaced coherent state. The center of the distribution is located at $(\langle \mathbf{x} \rangle, \langle \mathbf{p} \rangle)$. (c) A state squeezed in $\mathbf{x}$ ($s < 1$). (d) A state squeezed in $\mathbf{p}$ ($s > 1$).

graphy. The unit of quantum information theory, the qubit, was also detailed. Finally, I gave some elements of quantum optics, to which the implementation of quantum cryptography is closely related.

For further reading, a good introduction to quantum information theory can be found in the book by Nielsen and Chuang [138] and in Preskill's lecture notes from the California Institute of Technology [142]. For a more general treatment of quantum mechanics, one can read the book by Greiner [73]. For quantum optics, I suggest the books by Scully and Zubairy [158] and by Walls and Milburn [180].

5

Cryptosystems based on quantum key distribution

In this section, I wish to put quantum key distribution (QKD) in the wider context of a cryptosystem. I shall discuss informally some aspects that are considered important. The questions I wish to answer here are: "What are the ingredients needed to make a QKD-based cryptosystem work? What services does it bring? What are its limitations?"

As I shall detail below, QKD may be used to provide the users with confidential communications. This can be achieved when we combine QKD and the one-time pad. For the quantum modulation, QKD needs a source of truly random numbers. Also, QKD requires a classical authenticated channel to work, so authentication plays an essential role. As a consequence, QKD must start with a secret key, making it a secret-key encryption scheme. I will also discuss what happens if classical cryptography is introduced in the system. Finally, I will describe the implementation of a simple cryptosystem on top of QKD.

5.1 A key distribution scheme

The first function of QKD is to distribute a secret key between two parties. The use of this key is outside the scope of this first section – the need for a secret key is omnipresent in cryptography.

As depicted in Fig. 5.1, QKD relies on a classical authenticated channel for sifting and secret-key distillation and on random bits for the modulation of quantum states. The key produced by QKD can be intended for encryption purposes – this will be discussed in Section 5.2 – but is also required by authentication. A part of the distributed key is used for authentication. When QKD is run for the first time, however, an initial secret key must be used instead.

The platform, that is, hardware, software and other practical devices and

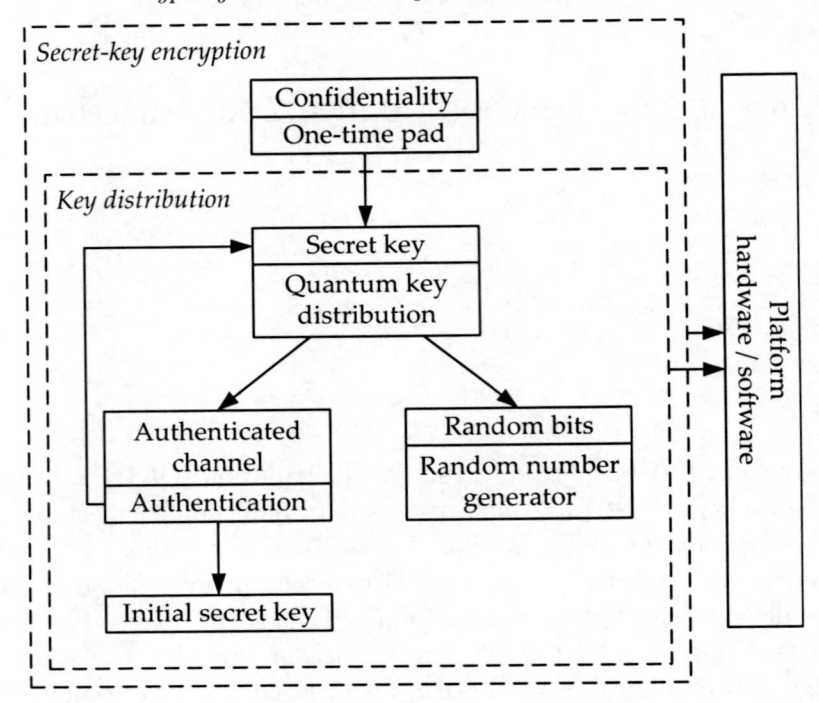

Fig. 5.1. The nominal case usage of QKD for key distribution and for secret-key encryption. An arrow represents a "depends on" relationship.

matters, is usually left outside of the cryptographic schemes and is assumed to be perfect. Nevertheless, it is explicitly displayed in Fig. 5.1 to remind the reader that security breaches can also come from this part of the system.

In the following sections, I will review the components of the key distribution scheme, including the classical authenticated channel and the source of random numbers.

5.1.1 The classical authenticated channel

For QKD to work, Alice and Bob need to talk over a classical authenticated channel. This is necessary for at least two reasons. First, the parties need to probe the quantum channel so as to have an estimation of the information leaked on it. For this, they need to compare samples of their transmissions and measurements. Second, the parties need to turn the sequence of exchanged random variables into a common secret key, so some communication is necessary to achieve this.

Authentication of classical messages is an essential ingredient for QKD to work. It allows the recipient of a message to detect whether that message was written by the person supposed and was not tampered with. Without the classical authenticated channel, the legitimate parties, Alice and Bob, would not be able to tell whether they are discussing together or with Eve pretending to be one of them.

The attack where Eve pretends to be Alice when talking to Bob and vice versa is called a *(wo)man-in-the-middle attack*. Failing to prevent this attack enables Eve to exchange a key k_A with Alice and another k_B with Bob. Then, when Alice sends a private message, Eve can decrypt it using k_A, re-encrypt it with k_B and forward it to Bob, or vice versa when Bob sends something to Alice. This is Eve's complete control of Alice and Bob's communications, and nothing suspicious seems to happen.

It is important to remind that QKD does not solve the key distribution problem without the need of a bootstrap key for authentication. At some point, Alice and Bob must be able to check they are talking to one another.

Authentication can be implicit or explicit. When implicit, the context unambiguously ensures that Alice can safely trust that she is talking to Bob and vice versa. For instance, they know each other, and personal features such as the voice, the facial expressions and some common personal memories ensure that they are talking to the right person. This is often insufficient, as discussed below.

Explicit authentication is when the messages sent over the classical channel are authenticated by some cryptographic technique such as a MAC – see Section 2.2. This is necessary when Alice and Bob are not continuously seeing each other or if they are just pieces of software. In such cases, Alice and Bob need to share a secret key prior to the first message they wish to authenticate. In the scope of QKD, it means that, although QKD allows two parties to exchange keys, it relies on authentication, so the first authentication key must be distributed by some other means. (We think here about symmetric authentication techniques – for a discussion on asymmetric techniques, please refer to Sections 5.3.1 and 5.3.2.) An obvious solution is for Alice and Bob to meet at least once and agree, face-to-face, on a randomly chosen key. Afterwards, they can use a QKD protocol to exchange a key, part of which can be used to authenticate the next run, and so on.

When implicit authentication is not enough

Although it is reasonable to assume as authenticated the communications in a face-to-face discussion of two people as close as Alice and Bob, one may have more doubts about what happens when they are talking through a telephone or a video conference system. Sure, as Alice and Bob discuss which bit parities differ (as in Cascade, see Section 8.3.2), they can interleave their discussion with jokes. Alice knows Bob's taste for blonde jokes, even if she does not approve it, while Bob cheerily enjoys Alice's endless repertoire of tasteful puns, so they can make sure they are not talking to each other's double.

But this has some limitations. As Eve comes to play with her all-powerful computer, she has no difficulty in recording the sound of Alice and Bob's voices saying "0" and "1" and replaying them at appropriate times when the two parties exchange bits. Now, Alice and Bob's jokes will be of no use, since Eve replaces the zeroes and ones of the legitimate parties with values of her own choice, and this does not disturb their hilarious interleaved discussions.

Of course, Alice and Bob can start thinking about a better way to authenticate their discussions. For instance, they can number their jokes from 0 to 1023 (i.e., 10-bit strings). After Alice and Bob have exchanged 10 bits, they stop and tell the corresponding joke. They can think of even more complex mechanisms, such as making the joke number depend on the time slot of the message to prevent Eve from re-using messages. But as Eve listens, she inevitably acquires the knowledge of a large subset of the 1024 jokes and starts attacking with the bits of her choice, thanks to her most advanced voice synthesizer. After some thinking, Alice and Bob will find their joke-authentication scheme not so funny anymore and will turn to something else.

This allows us to draw the following informal conclusions. Without any assumptions on Eve's capabilities, it is difficult to ensure authentication relying only on the context. If she is allowed to use any technique, Eve can fool Alice and Bob over the telephone or the video conference and less human-friendly systems such as a network connection. Then, some form of secret key is necessary, no matter if it is a binary string or a private memory (see also [59]). Furthermore, this key is progressively learned by Eve as the legitimate parties use it. Finally, it is necessary to make authentication depend on both the messages and the secret key.

Besides human face-to-face meeting, implicit authentication is not enough and should be replaced by explicit authentication.

Explicit authentication techniques

As explained in Section 2.2.2, there exist information-theoretic authentication schemes, that is, schemes without any computational assumptions. A well-studied technique for designing information-theoretic MACs is based on strongly universal families of hash functions, as proposed by Wegman and

Carter [182]. In this section, I will first define such a family of functions and then give an example of construction.

Definition 1 ([169, 182]) *Given two sets $\mathcal{A}$ and $\mathcal{B}$, a class $\mathcal{H}$ of functions $\mathcal{A} \to \mathcal{B}$ is $\epsilon/|\mathcal{B}|$-almost strongly 2-universal if the following two conditions are satisfied:*

- *for any $x_1 \in \mathcal{A}$ and any $y_1 \in \mathcal{B}$, the size of the set $\{h \in \mathcal{H} : h(x_1) = y_1\}$ is at most $|\mathcal{H}|/|\mathcal{B}|$;*
- *for any $x_1 \neq x_2 \in \mathcal{A}$ and any $y_1, y_2 \in \mathcal{B}$, the size of the set $\{h \in \mathcal{H} : h(x_1) = y_1 \wedge h(x_2) = y_2\}$ is at most $\epsilon|\mathcal{H}|/|\mathcal{B}|^2$.*

If the last condition is satisfied for $\epsilon = 1$, the class is simply called strongly 2-universal. *For simplicity, we assume that* strongly universal *means* strongly 2-universal.

In general (and even outside the context of cryptography), a hash function is a function that maps a larger set to a smaller one with the property that any two different inputs are likely to yield different outputs. In classical cryptography, cryptographic hash functions are often used in combination with signature schemes to act as message digests – see Section 2.3.2.

In contrast to using a single well-defined hash functions as a message digest, our purposes require us to use a family of hash functions. The choice of the particular hash function to use is kept secret by the legitimate parties. The construction of MACs for quantum cryptography shares some similarity with keyed hash functions such as HMAC [7]. Unlike HMAC, however, the security of our construction comes from the properties of the family itself (i.e., strong universality), not very much from the properties of the individual hash functions, which can be very simple.

Given an $\epsilon/|\mathcal{B}|$-almost strongly universal family, the MAC is computed as

$$\text{MAC} = h_K(m),$$

with m the message to authenticate and K the secret key shared by the legitimate parties. Hence, a hash function is randomly and uniformly chosen among the family. The family must be able to process $a = |m|$ bits as input (i.e., $\mathcal{A} = \{0,1\}^a$) and give $b = |\text{MAC}|$ bits as output (i.e., $\mathcal{B} = \{0,1\}^b$).

As for the one-time pad, it is necessary to use a fresh key for every message to be authenticated. It follows that the family size $|\mathcal{H}|$ should be as small as possible: every MAC computation requires $\log |\mathcal{H}|$ key bits to identify the hash function within the family.

In terms of security, the use of an $\epsilon/|\mathcal{B}|$-almost strongly universal family of hash functions allows the MAC to have impersonation probability $1/|\mathcal{B}|$

and substitution probability upper bounded by $\epsilon/|\mathcal{B}|$ [169, 182]. No matter how powerful Eve's computer is and no matter how cryptanalysis evolves over the years, she will not be able to deceive the legitimate parties, except with an arbitrarily small probability.

An example of a strongly universal family of hash functions is given next.

Example 1 *Let $\mathcal{A} = \mathrm{GF}(2^a)$ and $\mathcal{B} = \{0,1\}^b$. Let $h_{c,d}(x)$ be defined as the first b bits of the affine function $cx + d$ in a polynomial representation of $\mathrm{GF}(2^a)$. The set*

$$\mathcal{H}^{(1)}_{\mathrm{GF}(2^a) \to \{0,1\}^b} = \{h_{c,d} \,:\, c, d \in \mathrm{GF}(2^a)\}$$

is a strongly universal family of hash functions [182]. (Note that the location of the b extracted bits does not matter for the family to be strongly universal.)

The strong universality of the family in Example 1 is easy to prove, as only one affine function can take prescribed values at two distinct abscissas. Note that actually only the first b bits of d influence the result. Hence, we do not need to specify all the a bits of d and we can, instead, perform the addition after extracting the b bits. The modified family is denoted $\mathcal{H}^{(2)}_{\mathrm{GF}(2^a) \to \{0,1\}^b}$ and is also strongly universal.

If directly used as a MAC, this family is not very efficient: it requires $a+b$ key bits (i.e., a bits for c and b bits for d), more than the message itself. As sifting and secret-key distillation need to be authenticated, the exchanged messages are expected to be larger than the size of the produced key, hence consuming up all the secret key bits.

The output size b of the family is essentially determined by the deception probability that one wants to tolerate and can be considered constant. E.g., the value $b = 256$ bits seems fairly conservative. On the other hand, one should be easily able to choose the input size a while preventing the family size from growing too quickly with a. This is the aim of the next construction.

Let us describe a construction proposed in [182]. For this, let

$$s = b + \log \log a$$

and $\mathcal{H}_0$ be a strongly universal family of hash functions that maps $2s$ bits onto s bits, for instance the family of Example 1. The family $\mathcal{H}_0$ is used as a primitive to build a $2/|\mathcal{B}|$-almost strongly universal family of hash functions from a bits to b bits.

First, we cut the message into $\lceil a/2s \rceil$ blocks of $2s$ bits each. If a is not a multiple of $2s$, some padding is applied to the last block. Each block is processed through a function $h_{0,1}$ randomly chosen from $\mathcal{H}_0$. The outputs are concatenated, yielding a string of about half of the initial size. Then, the resulting string is again cut into blocks of $2s$ bits, and each block is processed through $h_{0,2} \in \mathcal{H}_0$. The process is iterated until the bit string has length s. Finally, the first b bits of the s-bit string are taken as the result.

Notice that the same hash function $h_{0,i}$ is used to process all the blocks at a given iteration i. However, each hash function $h_{0,i}$ has to be chosen independently for each iteration. The description of a hash function within $\mathcal{H}$ thus consumes $\log|\mathcal{H}_0|$ bits times the number of iterations, which is roughly equal to $\lceil \log a - \log s \rceil$. Assuming the choice of the $\mathcal{H}_0 = \mathcal{H}^{(2)}_{\mathrm{GF}(2^{2s}) \to \{0,1\}^s}$ family, the number of necessary key bits grows as $3(b + \log\log a)\log a$ with the message size a and MAC size b.

Many variants are possible and other families of hash functions can be constructed – see for instance [135, 170] and the references therein.

5.1.2 The source of random numbers

To meet the conditions of perfect secrecy, the entropy of the key K must be maximal, see Section 3.6. The value of the key K is essentially determined by Alice's modulation. So, it is important that Alice's modulation is random and unpredictable.

Alice needs to modulate her quantum states in a random way. What does "random" mean in this context? It is not the purpose of this section to discuss the philosophy behind randomness – for a discussion on the definition of randomness, please refer to [96] and the references therein. Simply stated, we require that the random numbers are unpredictable to any potential eavesdropper. Since QKD is intended to be secure against an eavesdropper with unbounded computational resources, we must rule out pseudo-random techniques, that is, deterministic algorithms that generate random-looking bits.

Pseudo-random techniques are based on the algorithmic expansion of a small number of random bits, called the *seed*, into a much larger number of pseudo-random bits. Following Kerckhoffs' principle [95] and thus assuming that the algorithm is publicly known, the entropy of the generated key is not greater than the entropy of the seed S, $H(K) \leq H(S) \ll l$.

We can instead try to use physical principles to produce random bits. Examples of physical random sources are noise in resistors, delays of radioactive decays and quantum noise.

There currently exist several devices that generate random numbers based on electronic noise. For instance, the QNG Model J1000KU™ of ComScire

[43] uses a combination of thermal or Johnson noise and amplifier noise, which can produce random bits at a net rate of 1 Mbps through an USB interface. Another example is the R200-USB^{TM} serie from Protego [143], which produces up to 2 Mbps. By using a faster interface or by embedding the technology in a PC, higher rates are probably not very difficult to achieve.

Random bits can also be generated using quantum noise. In principle, measuring the state $2^{-1/2}(|0\rangle + |1\rangle)$ in the $\{|0\rangle, |1\rangle\}$ basis gives a random bit. This principle is implemented in Quantis^{TM} from id Quantique using single photon detectors [90]. The de-biasing of the generated bits is ensured by Peres' method [141]. They achieve a rate of 4 Mbps in a device of the size of a matchbox. A model that combines four such devices can give a rate of 16 Mbps.

Regarding radioactive decays, we can mention HotBits [179], which generates random bits by timing successive pairs of radioactive decays detected by a Geiger–Müller tube. Unfortunately, the rate of 240 bps is quite low for our purposes.

Clearly, the use of quantum randomness for QKD is ideal. However, although one may argue that noise in electronic components is ruled by deterministic equations, it is still acceptable. In this case, the generation of random bits may have the same drawback as pseudo-random generators: if an eavesdropper can simulate the electronic circuit of Alice's random number generator, the entropy of the generated bits is upper bounded by the initial state of the simulated physical system. Yet, the main difference with algorithmic pseudo-random generator is that in this case the entropy of the initial state is huge. Let us take as a reference Avogadro's number, $N_{\text{Avogadro}} \approx 6.022 \times 10^{23}$. If the entropy of the initial state is of the order of Avogadro's number, $H(S) = N_{\text{Avogadro}}$, Alice can generate random numbers for 19 million years at a 1 Gbps rate. Hence, we can safely assume that $H(S) \gg l \approx H(K)$ for practical values of l and thus both quantum noise and electronic noise are sufficiently good sources of randomness for our purposes.

5.2 A secret-key encryption scheme

As depicted in Fig. 5.1, the scheme we consider in this section is the exchange of a key between Alice and Bob using QKD, the public discussions being authenticated with an information-theoretic MAC and the encryption being performed with one-time pad using the exchanged key.

We have shown that this scheme requires a short initial secret key as bootstrap. It thus fulfills the same function as a classical secret-key encryption scheme, namely the encryption of long streams of data with a much smaller key to start with. The difference between this scheme and a scheme based on classical ciphers is where the security comes from.

The quantum channel can be attacked, and this is obviously the subject of many papers in QKD. Even if Eve has an infinitely powerful technology (including computational resources) to do so, Alice and Bob can have some statistical guarantee on the limited impact of her attacks. For instance, they make sure the eavesdropping does not leak more than ϵ bits of information on the key, otherwise with probability p_Q.

These attacks show some resistance against technological advances. Unless long-term reliable quantum memories exist at this time, Eve must use today's technology to attack the key exchanged today. This remark may be important for long-term secrets and is quite unique to quantum key distribution; see also Section 5.3.2.

Of course, there are not only attacks on the quantum part of the process. As explained earlier, there is a non-zero but arbitrarily small probability p_{Auth} that Eve succeeds in fooling Alice and Bob's authentication and in taking part in the public discussion. But there are many other sources of failure, for technological and human reasons. For instance, Eve breaking into Alice's office to steal the key is very well allowed by the laws of quantum mechanics.

Also, an insecure platform may open security holes. The quantum source and detector may not behave exactly as modeled. The hardware itself can also leak information before sending a quantum state or after measuring one. As an example, Eve may use the quantum channel to send some photons back to Alice such that the reflection of these photons on her apparatus gives information on the key [65]. The plaintext data may still be recoverable from Alice's hard disk, even if she erases it, because of the hysteresis of the electromagnetic support. Or the legitimate parties' hardware may otherwise simply leak secrets through electrical or electromagnetic signals as exploited by implementation attacks of classical cryptographic devices [94, 97]. And these implementation-related leakages are especially threatening with QKD, as it requires fairly sophisticated hardware.

To sum up, an encryption system based on QKD provides the same service as a secret-key classical cipher, but with a different support for security – and only a small part of it, if you take the complete system into account. While classical ciphers essentially assume that no one can design an algorithm or a computer that can break the cipher in a reasonable time frame, QKD-based encryption relies on strong physical principles (i.e., quantum mechanics) but is also prone to surprises with regard to its implementation, which may not behave exactly as in theory.

5.2.1 What about public-key cryptography?

If this QKD-based cryptosystem compares to secret-key encryption, what about public-key cryptography? In fact, QKD and public-key cryptography are fairly different matters.

Like classical ciphers, public-key primitives such as RSA or Diffie–Hellman rely on computational assumptions, that is, on the difficulty of solving some mathematical problems. In the case of public-key cryptography, such problems are often factoring large numbers or calculating discrete logarithms. As for classical ciphers, such assumptions may some day become invalid because, e.g., fast algorithms are found, computers become really fast or a combination of both [162].

There are, however, some major differences between public-key encryption and secret-key classical ciphers. Because public keys are by definition known to many people, there is no need to exchange them secretly. In contrast, both secret-key classical ciphers and the cryptosystem based on QKD of this section need to start with a key that was exchanged between Alice and Bob by some secure means. Another consequence of the usage of public keys is that anybody can encrypt messages to a given recipient. Neither secret-key classical ciphers nor QKD-based cryptosystems can provide such a service.

So the function of public-key encryption cannot really be compared to that of the QKD-based cryptosystem of this section.

By contradiction, no public-key encryption can be free of computational assumptions. If Eve can encrypt any message m to Bob using his public key K_B, she can, using her unbounded computational resources, prepare a list of all possible messages together with their encrypted versions: $\{(m, K_B(m))\}$. When Alice sends an encrypted message c to Bob, Eve can look up her list and find m such that $c = K_B(m)$. This aspect is formalized in [118]. Note that we assumed a deterministic encryption scheme, but the argument can be easily adapted to the case of probabilistic encryption such as, e.g., the Goldwasser–Micali scheme [67].

Public-key cryptography is often used in combination with secret-key cryptography, with the exchange of an intermediate secret key. For instance, the function of Diffie–Hellman's protocol is to provide two parties with a secret key that can later be used for encryption using a secret-key cipher. Because of its relatively slow speed, RSA is often used to encrypt a randomly chosen secret key, later used with a secret-key cipher, rather than

5.3 Combining quantum and classical cryptography

to encrypt the whole message itself. So, this key-exchange aspect of public-key cryptography looks similar to QKD. However, the fact that intermediate keys are useful in both cases does not change the functional differences described above and thus does not seem relevant.

5.3 Combining quantum and classical cryptography

In this section, I wish to show informally that QKD has quite unique properties, which classical cryptography does not have, and that combining QKD and classical ciphers is, in general, not such a good idea.

5.3.1 Signatures for authentication

To overcome the need for a bootstrap key, some people suggest using public-key (PK) signatures to authenticate the messages over the classical channel – see, e.g., [57, 58]. That is, we replace the secret-key authentication primitive of Fig. 5.1 with a PK signature scheme.

Using a similar argument as above, PK signatures cannot be free of computational assumptions. With the traditional argument that a chain is as strong as its weakest link, breaking a QKD scheme with PK signatures now reduces to breaking the signature scheme.

Given some interactivity of the classical messages being signed, one may argue that the attacker has to break the PK signature scheme in a fairly short time. If Alice and Bob exchange their public keys just before running the QKD protocol, only then can Eve start attacking the PK signature scheme, and she must be done when the first classical message is sent.

So, the situation may seem a little less pessimistic, but actually it is not the case. Public keys may be distributed in the clear, but one must be able to check the association between the public key and its owner. Otherwise, Eve may as well give her public key to Alice and Bob instead of the legitimate ones. Public key infrastructure associates a key with a certificate, which provides a signature from a certification authority (CA). So, if Eve does not have much time to break Alice and Bob's signature scheme, she can instead take more time and break a CA's signature scheme, so as to sneak her own public key instead of Alice's and Bob's, without being noticed.

Using a public-key scheme removes the need of an initial secret key and thereby may give the impression of requiring no bootstrap phase. The bootstrap phase with a PK scheme is the fact that some CA must verify Alice's and Bob's identities, which may have be as expensive as a face-to-face meeting to exchange an initial secret key.

5.3.2 Strong forward secrecy

In classical cryptography, some cryptosystems use session keys, that is, short-term keys that are derived from one or more long-term keys. A fraction of such systems has the property of achieving *perfect forward secrecy* [81, 127], that is, the guarantee that a compromised long-term key does not compromise past session keys. In the context of QKD, a similar property is verified.

QKD can provide *strong forward secrecy* in the following sense: the compromise of an authentication key after being used does not compromise the secrecy of both past and future messages encrypted with QKD-generated keys [140].

Once some key material has been generated by QKD, it can be used to authenticate a following round of QKD to produce further key material. So, if Alice and Bob use computationally secure classical authentication to start the first run of QKD, Eve has to break it before or during its execution. If Alice and Bob are confident that their classical authentication was not broken at the time they used it, all the confidential messages encrypted with QKD-produced keys remain secret forever. As time goes on, Eve may then become able to break the authentication scheme and even to recover the authentication key material they used, but the opportunity for her breaking into Alice and Bob's confidentiality is gone.

With classical cryptography only, one cannot achieve strong forward secrecy: assuming that Eve listens to all the communications, her ability to break the system means that all future communications can be broken as well. Perfect forward secrecy only guarantees that past communications remain secure, whereas strong forward secrecy keeps both past and future communications secure.

Strong forward secrecy is one of the strong points of QKD, as it can ensure the confidentiality of long-term secrets. For instance, if a company wants to archive long-term secret data to a remote location, the use of QKD for this purpose implies that the eavesdropper has to use today's technology to break today's transmission.

Although this concept works for both information-theoretic authentication and for computationally-secure classical authentication, it is quite interesting in combination with authentication using public-key signatures: Since the encryption-scheme can be started with public keys, it becomes a public-key encryption scheme. However, this has limitations that are now discussed.

First, it does not contradict the conclusion of Section 5.3.1. If at some point Eve has enough power to break the public-key signature system, no further new pair of communicating parties can start using QKD with this PK signature system for authentication. Or stated otherwise, the computational assumptions of PK signatures have an impact on the bootstrap phase but not on established pairs of users. For instance, Alicia and Bobby, the daughter and son of Alice and Bob, may have some trouble using this technique in 2047, if by then public-key schemes have become insecure.

Second, long-term security of the entire system is not obvious to maintain. As time goes on, Eve has had more and more time to recover the certification authority's private key, and when she eventually succeeds, she can sneak a fake certificate into the system and break the authentication of any initiating QKD-based system. If the CA wants to refresh the certification hierarchy, it may be too late: who can still trust it? Note that maybe the CA can provide new keys and new certificates very often, so as to limit the impact of this, but with an extra cost.

Finally, denial-of-service attacks can be mounted so as to force Alice and Bob to waste their key material. Since the key for one-time pad or authentication cannot be reused, messages corrupted or discarded by Eve use up secret key bits. If this happens at some point where the PK signature system is broken, Alice and Bob have no choice but to restart from the bootstrap phase, either with a stronger PK signature system, in which case they have to trust a new CA public key, or with information-theoretic authentication, for which they have to exchange a secret key by hand.

5.3.3 Classical ciphers for encryption

In implementations of QKD based on photon detectors, the net secret key bit rate R is fairly low, namely expressed in kilobits per second [192]. To overcome this, it is tempting to combine the security of QKD and the high rate of classical cryptography in the following way: Alice and Bob use a classical secret-key cipher to encrypt their messages, and the key used in the cipher is changed as often as possible using the quantum-exchanged key. That is, we trade the one-time pad for a classical (block or stream) cipher in Fig. 5.1.

This combination is suggested in many papers, e.g., [57, 58, 86, 87], and used in some commercial products.

With such a combination, breaking the encryption does not necessitate breaking both the QKD part and the classical cipher: either one of them is sufficient. So, roughly speaking, the security of the combined system cannot be greater than either of the constituting parts. In particular, the combined system is not more secure than the classical cipher alone. Long-term security and resistance against technological advances cannot be guaranteed any more: one can intercept Alice and Bob's encrypted messages and hope that a fast enough computer will be designed in the future to break the classical encryption.

Things can be even worse if one takes possible implementation flaws or breaches in the platform. In this case, the classical–quantum cryptography combination may even be potentially less secure than a classical cipher alone, as the platform becomes wider and potential security problems from the QKD hardware come on top of any potential implementation flaw in the classical scheme.

Sometimes, it is argued that changing the secret key of a classical cipher increases its security. Let us further investigate this idea.

In some commercial products, one can change the 256-bit AES key 100 times per second; we thus assume a QKD net secret key rate of 25.6 Kbps. The encrypted link, on the other hand, is a regular network connection, and let us assume that the data rate is 100 Mbps. With these settings, one 256-bit key is used to encrypt 1 Mbits ($\approx 2^{20}$ bits) using AES.

If one trusts the chosen classical cipher, why bother changing the key so often? It certainly does not diminish its security, but the improvement may not be significant or even quantifiable.

For AES, the majority of the cryptographic community believe that there are no attacks faster than exhaustive search at this time. They also believe no significant attacks exist on Triple-DES (excluding properties inherited from DES and from its short block size). So, it is impossible today to determine whether changing the key faster brings a significant improvement. If we look at what would be possible to attack with DES (not even Triple-DES or AES), changing the key even every few hours (instead of 100 times per second) would not change the security as known today. Linear attacks against DES can be mounted with a theoretical complexity of 2^{50} DES operations if 2^{44} known bits (i.e., 2^{38} blocks of 64 bits each) are encrypted with the same key [127]. With regards to this attack, changing a DES key every 2^{40} bits would be enough.

Furthermore, if the assumption is that a block cipher remains secure as long as it does not encrypt more than b blocks with the same key, then by contradiction one can construct a classical encryption scheme that safely encrypts a much larger number of bits. The stream of data is cut into chunks of b blocks, and each chunk i is encrypted with the session key $K_1(i)$. If Alice and Bob agreed on K_2, Alice can encrypt the b session keys $K_1(i)$, $i = 1 \ldots b$ with K_2 and send them to Bob. This way, we can safely encrypt

b^2 blocks, and of course this idea may be further repeated with t key levels to achieve the encryption of b^t blocks of plaintext.

So, if changing the key often does bring additional security, the classical–quantum cryptography combination is not more secure than a classical scheme with frequent session key changes.

Finally, some recent attacks concentrate on algebraic aspects of block ciphers [45, 131] and do not require many pairs of plaintext and ciphertext. Not going into the debate of whether they do work on AES or Triple-DES, the principle of these attacks remains unchanged if one changes the key more often.

As a conclusion, combining QKD and a classical cipher, in general, produces the least security of the combined systems. This conclusion remains valid even if changing the key often would improve security, a fact that cannot be proven or denied given the current research.

5.4 Implementation of a QKD-based cryptosystem

In this section, we wish to specify a simple yet concrete cryptosystem based on QKD. The goal is for Alice and Bob to obtain a usable key from the QKD-distributed key elements, assuming they agreed on suitable authentication and secret-key distillation protocols. I describe how authentication and secret-key distillation can interact in practice, when used, for instance, over a regular TCP/IP connection. There exist a variety of ways to devise such a cryptosystem, of which I propose an example.

Let us assume that the Alice–Bob convention is related to the quantum transmission: Alice sends quantum states to Bob. In this section, we use another naming convention. We call Claude (client-side) the party who will be sequencing the classical protocols and sending requests to the other party, called Dominique (server-side). Secret-key distillation cannot be fully symmetric; one given party has to start the classical transmission. Linking the two conventions allows us to assign Claude=Alice and Dominique=Bob, or Claude=Bob and Dominique=Alice. Note that in French the names Claude and Dominique are equally well used for men and women, hence making both assignments valid.

5.4.1 Pool of valid keys

On each side, Claude and Dominique maintain a pool of secret keys, that is, a list of bit strings associated with names. E.g., the pool is a set such as:

{(bootstrap, 1100101011001001 ...),
  (block1, 0110001010101101 ...), (block2, 1000011011000111 ...), ... }.

Such keys are either produced by QKD or given at startup for the bootstrap phase, i.e., to authenticate the first run of QKD.

We call a *valid key* a key that was correctly distilled with a QKD protocol (or given at startup) and that was not yet used to encrypt or authenticate any message. Since we wish to use the one-time pad for encryption and an information-theoretic MAC, any used key cannot be re-used and must be excluded from the pool of valid keys.

Validity is not symmetric in the two parties; a key can be valid for Claude but not valid for Dominique. Desynchronization can happen in practice. For instance, an encrypted message sent by Claude may be lost on its way to Dominique. Upon sending the message, Claude invalidates the key used to encrypt it. Dominique, however, is not aware that this key was used since she never receives the message, and so she does not invalidate it.

The key naming convention should of course be identical on both Claude's and Dominique's sides. We can imagine that the produced keys can have a name with a number increasing with each run of the QKD protocol.

To avoid wasting key bits, we wish to be able to access individual subsets of keys. The reason for this is that we may only need to encrypt a 1 Kb-long message while the keys produced by QKD are 1 Mb-long. Also, authentication requires a much smaller key than the one produced by QKD. When requesting a key, one should specify the offset and the number of bits needed within a key. For instance, the string made of the t consecutive bits of key with name key starting at offset o is written $key(o, t)$. The full name of a l-bit key produced by QKD is $key(0, l)$.

Accordingly, the invalidation of the key should be done only for the key part used. For instance, let the key $(key(0, l), k'k)$ be present in the pool of valid keys, with k' a l'-bit string and k a $l - l'$-bit string. After a request to invalidate $key(0, l')$, the pool contains $(key(l', l-l'), k)$. For simplicity and to avoid fragmentation, we enforce the rule that the key parts are invalidated from lower indexes on.

5.4.2 Layers and protocols overview

For simplicity, we assume that we have access to a socket connection (or similar) over a network (e.g., a TCP/IP connection over the Internet). This is the public classical channel, which is not yet authenticated: it is our responsibility to ensure authentication. In networking terminology, we use the application layer of the network connection. For our purposes, we subdivide the application layer into the following:

- The *command–response layer* must allow Claude to send commands to the other party, Dominique, who can then respond. Claude must be able to send any message in $\{0,1\}^*$, the set of finite-length binary strings, correctly delimited, so that Dominique can receive it without ambiguity. Dominique must respond to Claude's command before s/he can send further commands.
- The *authenticated layer* comes above the command–response layer. Application messages must be enclosed in a larger message construction, so as to be able to send commands and receive responses, possibly either or both of them being authenticated.
- Using the authenticated layer, the *block secret-key distillation protocol* (BSKDP) is a protocol that processes one run of the QKD protocol. It starts when the QKD has finished producing a block of raw key elements, and produces a block of secret key bits.
- Using the BSKDP, the *continuous secret-key distillation protocol* (CS-KDP) provides continuously new secret keys, using a part of the previously distilled secret key for authentication of the current block.

5.4.3 The command-response layer

The command-response layer must ensure that the receiver can determine the message boundaries, since the underlying socket only provides an undelimited stream of bytes.

Concretely, there exist many freely available libraries that readily implement some form of command and response passing. For instance, the command-response layer over a socket connection can be done using the ACE™ library [152].

5.4.4 The authenticated layer

Authentication is essential for QKD to work, and we will assume here the use of information-theoretic MACs as described in Section 5.1.1. Remember that the key material used for authenticating one message may not be reused for the next message. Since the number of key bits used depends less than proportionally on the message length, it is usually more interesting to authenticate a long message than several small ones. For this reason, we create authenticated sessions, that is, sets of consecutive commands or responses that are authenticated together.

We separate the authentication of commands (sent from Claude to Dominique) from the authentication of commands and responses together.

The sender, Claude, can decide to enclose a set of command messages in an authenticated session. The corresponding authentication code is called a CMAC (command message authentication code). Claude starts a CMAC session by using a special indication in a message that says that such a session begins. Upon CMAC session end, Claude must give the MAC of all the commands in the CMAC session. The CMAC is computed from the key and from the command messages in the session, properly delimited to avoid ambiguities, $C_{\mathrm{MAC}} = h_K(c_1|c_2|\ldots|c_n)$ with K the key and c_i the commands sent by Claude. A correct CMAC gives Dominique the certainty that s/he received commands from Claude but no one else.

Likewise, Claude can also request Dominique to send authenticated responses. Claude can request a CRMAC (command-response message authentication code) session to start and to end. Upon CRMAC session end, Dominique must send a response plus the MAC of the entire CRMAC session. The CRMAC is computed from the key and from the command messages and responses in the session, $\mathrm{CR}_{\mathrm{MAC}} = h_K(c_1|r_1|c_2|r_2|\ldots|c_n|r_n)$, with r_i the responses from Dominique. A correct CRMAC gives Claude the certainty that Dominique received the correct commands and that his/her responses were not tampered with.

The asymmetry between the CMAC and the CRMAC sessions is necessary. If the CRMAC covered only the response messages, Claude could not tell if Dominique was responding to the proper commands. We need at least one party who can tell if the entire dialog is consistent. Since Claude is in charge of sequencing the protocol, it seems fairly natural to give him/her this power in the CRMAC verification.

We assume that Claude decides when either the CMAC or the CRMAC session begins or ends. Deciding when to start or stop authentication is not part of the authenticated layer but part of the layers or protocols above. It is however the purpose of the authenticated layer to mark messages in such a way that the recipient of a message can determine whether the message is inside or outside a session and when a session begins or ends.

At this point, we consider that CMAC and CRMAC sessions can be completely independent. For each session, a message can be marked with any of these three symbols {in, last, out}. The CMAC session status is indicated by the variable $C_{\text{Session}} \in \{\text{in}, \text{last}, \text{out}\}$, whereas the CRMAC session status is indicated by $\text{CR}_{\text{Session}} \in \{\text{in}, \text{last}, \text{out}\}$.

The messages in a session, except the last one, are marked with in; the last message of a session is marked with last; a message outside a session (i.e., not authenticated) is marked with out. Thus, a message is in a session iff it is marked with in or with last. We say that a message begins a session when it is in a session and when the previous message was marked either with last or with out. We say that a message ends a session when it is marked with last.

As they are taken from the pool of valid keys, the keys to be used for the CMAC or CRMAC session must be identified. This is done by the variable $C_{\text{KeyID}} \in \{0,1\}^*$ (or $\text{CR}_{\text{KeyID}} \in \{0,1\}^*$) which indicates which key is used to calculate the CMAC (or CRMAC). This information must be given when the CMAC (or CRMAC) session begins.

Given a tuple $(c, C_{\text{Session}}, C_{\text{KeyID}}, C_{\text{MAC}}, \text{CR}_{\text{Session}}, \text{CR}_{\text{KeyID}})$, the authenticated layer encodes it as a binary message c', where c is an upper-layer command message, $C_{\text{MAC}} \in \{0,1\}^*$ is the CMAC of the session when $C_{\text{Session}} = \text{last}$ and c' is the command message to send via the command–response layer. Dominique must, of course, be able to recover the elements of the tuple from c'.

The authenticated layer also encodes the pairs $(r, \text{CR}_{\text{MAC}})$ to binary messages r', where r is an upper-layer response message, $\text{CR}_{\text{MAC}} \in \{0,1\}^*$ is the CRMAC of the session when $\text{CR}_{\text{Session}} = \text{last}$ and r' is the response message to send via the command–response layer. Likewise, Claude must, of course, be able to recover the elements of the pair from r'.

We must also specify how the validity of the CMAC/CRMAC sessions are processed. It is assumed that Claude always chooses a key ID that s/he considers as valid. So, a CMAC (or CRMAC) session is considered as valid iff the following conditions are satisfied:

- Dominique (or Claude) receives a CMAC (or CRMAC) from the other party that is equal to the one calculated internally;
- the requested key ID is known and the corresponding key is considered as valid by Dominique.

The fact that the CMAC/CRMAC sessions are valid or not does not influence the authenticated layer. It is only a piece of information that is accessible to the upper layers and protocols, which have to react appropriately to all situations. For instance, receiving a wrong CMAC or CRMAC during secret-key distillation means that the distilled key cannot be considered as valid.

5.4.5 The block secret-key distillation protocol

Let us now describe a list of steps that the implemented protocol must take to distill a key from a block of raw key elements produced by QKD. This part of the protocol relies on the authenticated layer, hence we have to specify the values of C_{Session} and of $\text{CR}_{\text{Session}}$. Unless otherwise specified, $C_{\text{Session}} = \text{CR}_{\text{Session}} = \text{out}$.

- ($\text{CR}_{\text{Session}} = \text{in}$, *CRMAC session begins*) Claude sends quantum transmission parameters (unique block number, physical parameters and any other variable the other party would need), together with randomly selected samples and their indexes. Dominique compares these with his/her own corresponding samples and estimates the eavesdropping parameters (e.g., an upper bound on the number of bits known to Eve due to eavesdropping on the quantum channel), which s/he memorizes. Depending on the correlations between the transmitted variables and the measurements (e.g., the bit error rate), Dominique specifies which protocol to use for reconciliation and its parameters. Dominique responds with the reconciliation specifications.

- ($\text{CR}_{\text{Session}} = \text{in}$, *CRMAC session continues*) Claude sends the first reconciliation message (e.g., some parity bits from Cascade or the syndrome of some error-correcting code, see Chapter 8). Dominique takes it into account, and possibly responds with his/her own reconciliation message (e.g., his/her parity bits from Cascade). Claude takes it into account.

- ($\text{CR}_{\text{Session}} = \text{in}$, *CRMAC session continues*) If the reconciliation is interactive and as long as it is necessary, Claude sends reconciliation messages, to which Dominique responds.

- ($\text{CR}_{\text{Session}} = \text{last}$, *CRMAC session ends*) Claude sends a request to Dominique to proceed with privacy amplification. Dominique chooses the privacy amplification specifications (e.g., hash function to use, final key size, see Chapter 6) depending on the eavesdropping parameters determined in the first step and on the number of reconciliation bits exchanged. Dominique calculates the final key by performing privacy amplification and responds with the privacy amplification specifications. (As the CRMAC session ends, the CRMAC is also sent by Dominique at the authenticated layer level.)

After this part of the protocol, Claude can consider the protocol as successful – and the distilled key as valid – if s/he receives the correct CRMAC from Dominique. In all other cases (incorrect CRMAC or none at all, zero final key size, etc.), Claude must consider the current run of the BSKDP as failed and may not add the produced key to the pool of valid keys.

If Claude receives a correct CRMAC from Dominique, s/he knows that Dominique responded to the correct commands and that the entire dialog is authenticated. The CRMAC ensures that Dominique correctly received the transmission parameters and samples, so that Claude can trust the estimation of the eavesdropping level and thus of the final key size s/he receives from Dominique. Also, the reconciliation protocol is guaranteed to have been performed correctly. The final key can thus be considered by Claude as equal and secret and thus put in the pool of valid keys. In particular, Eve could not mount a (wo)man-in-the-middle attack.

Dominique, on the other hand, does not yet know if Claude received correct messages from him/her; Dominique cannot yet conclude that s/he holds a valid key as at this point s/he might have received messages from someone else. To remedy this, we add the following step:

- (C_{Session} = last, *CMAC session begins and ends*) If the CRMAC from Dominique is correct, Claude assesses the validity of the transmission by sending a confirmation message with the block number. If the CMAC is correct, Dominique considers the distilled key as valid. S/He responds with an empty message.

As mentioned in Section 5.4.1, it is possible that one party can consider the key as valid, but not the other. This is also the case if this last message is lost.

Variants of the BSKDP are possible. For instance, Claude can request Dominique to send his/her randomly selected samples. Using the CRMAC, Claude is guaranteed to receive them from Dominique and not from Eve. Claude can estimate the eavesdropping level and final key size and send it to Dominique. Also, the CMAC session can also include all the steps, not only the last one.

5.4.6 The continuous secret-key distillation protocol

The continuous secret-key distillation protocol is simply a sequence of QKD transmissions, each followed by a block secret-key distillation protocol, possibly interleaved with the transmission of confidential messages. The CSKDP must be bootstrapped by an initial key, to authenticate the first BSKDP.

The initial key should be manually inserted in the system and automatically considered as valid.

Before each run, Claude and Dominique can synchronize their pool of valid keys so as to determine which key elements they still have in common. This synchronization is not a critical operation, but it may help the overall protocol to be more robust. It should avoid Claude requesting Dominique to use keys that s/he does not have (e.g., if for some reason one considers the key as valid and the other one does not) and thus waste precious key bits.

A (wo)man-in-the-middle attack against the synchronization procedure may make Alice and Bob incorrectly exclude keys they considered valid. This could be used by Eve to mount a denial-of-service attack, that is, to use up all the QKD-distributed key bits. On the other hand, the synchronization procedure is not critical in the sense that it will otherwise not compromise the security of the system: Eve cannot maliciously add her own key to the pool of valid keys. A possible trade-off is to authenticate synchronization with a computationally-secure classical MAC, which uses always the same key, only to prevent this denial-of-service attack.

5.5 Conclusion

In this chapter, I showed that QKD-based cryptosystems can be used to distribute usable secret keys given an access to suitable authentication techniques and to a source of random bits. The distributed keys can also be used to encrypt data with the one-time pad. I investigated the combination of QKD with computationally-secure cryptography techniques. Finally, I described a concrete cryptosystem based on QKD.

In the following chapters, we will see how one can distill a secret key, suitable for a QKD-based cryptosystem, from partially correlated and partially secret data.

6
General results on secret-key distillation

Secret-key distillation (SKD) is the technique used to convert some given random variables, shared among the legitimate parties and a potential eavesdropper Eve, into a secret key. Secret-key distillation is generally described as a protocol between the legitimate parties, who exchange messages over the public classical authenticated channel as a function of their key elements X and Y, aiming to agree on a common secret key K.

We assume here that the eavesdropper's knowledge can be modeled as a classical random variable. For more general assumptions on the eavesdropping techniques, please refer to Chapter 12.

The quantum transmission is assumed to be from Alice to Bob. In order to be able to specify another direction for secret-key distillation, we use the Claude-and-Dominique naming convention. Here, the random variables X and Y model the variables obtained by Claude and Dominique, respectively, using the quantum channel, and the random variable Z contains anything that Eve was able to infer from eavesdropping on the quantum channel. As detailed in Section 11.3.2, both assignments (Alice = Claude ∧ Bob = Dominique) and (Alice = Dominique ∧ Bob = Claude) are useful.

In this chapter, I first propose a general description of the reconciliation and privacy amplification approach. I then list the different characteristics that a SKD protocol can have. And finally, I overview several important classes of known results and treat the specific case of SKD with continuous variables.

6.1 A two-step approach

As illustrated in Fig. 6.1, secret-key distillation protocols usually involve two steps: reconciliation and privacy amplification. First, reconciliation insures that both Claude and Dominique agree on a common string Ψ, that is not necessarily secret. Then, privacy amplification produces a secret key K from Ψ. Not all results discussed below need to approach distillation in two steps. However, this division is fairly natural from both practical and theoretical points of view.

With reconciliation as a first step, Claude must send Dominique some redundancy information, which we denote as M. The transmission of reconciliation information is monitored by Eve and must be taken into account in Eve's variable Z' for privacy amplification. The random variable Eve has access to is denoted as $Z' = (Z, M)$, where Z is the information eavesdropped on the quantum channel before the reconciliation.

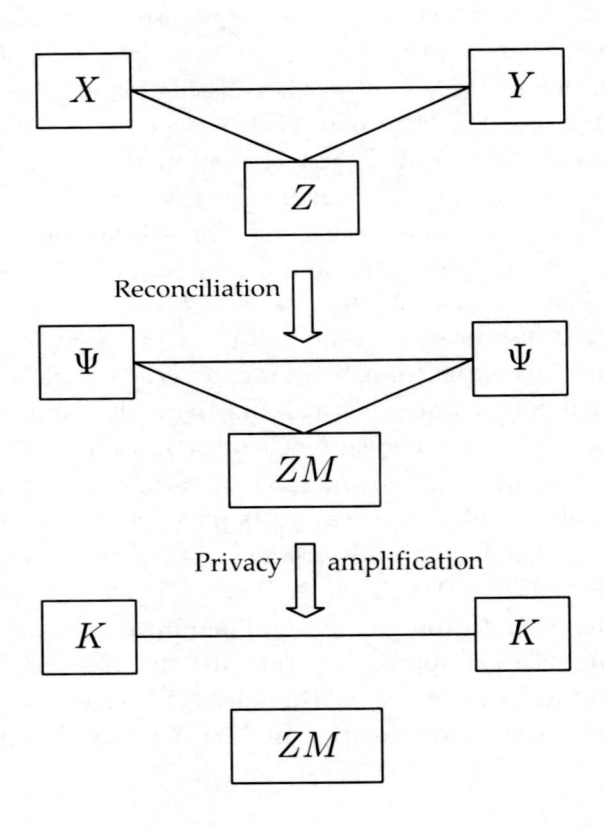

Fig. 6.1. Overview of the two-step approach for secret-key distillation: reconciliation and privacy amplification

6.2 Characteristics of distillation techniques

I shall now give an overview of the possible ways to perform secret-key distillation. These techniques have different characteristics, which are reviewed now.

First, the secret-key distillation can be either *one-shot* or *repetitive*. In the former, the SKD protocol has access to only one outcome (x, y) of the random variables X and Y – see also Fig. 6.2. It may model the result of one impulse sent over the quantum channel (where X and Y typically denote one-dimensional variables) or a complete run of a QKD protocol (where X and Y typically denote vector variables). In the repetitive setting, the SKD protocol has access to many independent outcomes $(x_1, x_2, \ldots, x_n, y_1, y_2, \ldots, y_n)$ of the same random variables – see also Fig. 6.3. This is suitable when a run of a QKD protocol can be modeled as having the same modulation distribution and when the eavesdropping follows the same strategy for every impulse sent.

Intuitively, the results regarding one-shot SKD have more of a worst-case flavor than for repetitive SKD. In the former, Claude and Dominique must be able to control the secrecy under any circumstances, whereas in the latter the law of large numbers allows them to consider only average circumstances.

Second, the public classical channel is assumed to be either *authenticated* or *unauthenticated*. In the former, it is the responsibility of the protocol user to insure that the public channel is authenticated. It is typically assumed that Claude and Dominique already share a short secret key before running the SKD protocol (see Section 5.1.1), so that they can authenticate the messages on top of the SKD protocol. In the latter, the SKD protocol ensures the authenticity of its messages, so that no assumption must be made on this aspect.

Third, the protocol can use either *one-way* or *two-way* communications. One-way communications are assumed from Claude to Dominique, so that the protocol is summarized into what s/he has to send. Two-way communications in general imply that one party has to wait for the other's message before sending anything.

Finally, the involved random variables may be either *discrete* or *continuous*.

The remainder of this chapter is organized as follows. I shall first review the authenticated discrete SKD results, starting from the one-shot distillation to the repetitive distillation. Then, I explain what happens in the case of unauthenticated SKD. Finally, I explain how some results can be generalized to continuous variables.

6.3 Authenticated one-shot secret-key distillation

In this section, I consider the case of authenticated SKD from a single outcome (x, y, z) of discrete random variables X, Y and Z, as illustrated in Fig. 6.2. We consider two-way communication, even though it also applies to one-way communication.

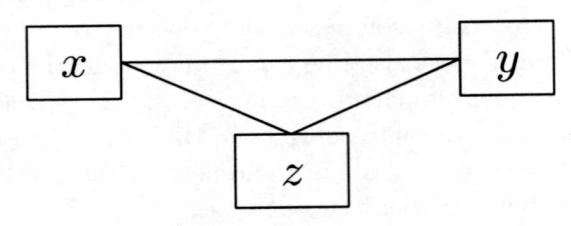

Fig. 6.2. In the one-shot setting, Claude, Dominique and Eve are assumed to have only one outcome (x, y, z) of the random variables X, Y and Z.

6.3.1 Privacy amplification with universal families of hash functions

Universal families of hash functions were created by Carter and Wegman for data storage and retrieval [38] and then later extended to authentication and set equality purposes [182]. Privacy amplification with such families of hash functions was first proposed by Bennett, Brassard and Robert [11]. In their paper, they considered deterministic eavesdropping strategies. Then, Bennett, Brassard, Crépeau and Maurer [15] generalized to the case of probabilistic eavesdropping strategies.

To achieve SKD using privacy amplification with universal families of hash functions, we assume that Claude and Dominique perform reconciliation and agree on a common variable, Ψ, beforehand. For simplicity, we assume that $\Psi = X$, that is, Claude's value X serves as a reference intermediate key. The following can be adapted trivially if Ψ is determined otherwise.

A universal family of hash functions, as defined below, serves the purpose of uniformizing the resulting random variable.

Definition 2 ([38]) *Given two sets $\mathcal{A}$ and $\mathcal{B}$, a class $\mathcal{H}$ of functions $\mathcal{A} \to \mathcal{B}$ is $\epsilon/|\mathcal{B}|$-almost 2-universal if for any $x_1 \neq x_2 \in \mathcal{A}$, the size of the set $\{h \in \mathcal{H} : h(x_1) = h(x_2)\}$ is at most $\epsilon|\mathcal{H}|/|\mathcal{B}|$. If this condition is satisfied for $\epsilon = 1$, the class is simply called 2-universal. For simplicity, we assume that* universal *means* 2-universal.

Although using the standard notation for the "almost" parameter, we

6.3 Authenticated one-shot secret-key distillation

favor the use of an expression of ϵ that indicates the deviation from universality, independently of the output set size: $\epsilon = 1$ iff the family is universal. The ϵ usually used in the literature satisfies $\epsilon_{\text{literature}} = \epsilon_{\text{here}}/|\mathcal{B}|$.

Notice that we met *strongly* universal families of hash functions in Section 5.1.1 for authentication purposes. For privacy amplification, strong universality is not required, but plain universality is. Strong universality implies universality.

Given that Eve does not know too much (in a sense made precise in the theorem below) about the variable X, a hash function chosen at random gives a result on which Eve does not have any information, except with negligible probability. The following theorem is slightly generalized from the one of [15] in the simple fact that cases for which $\epsilon \neq 1$ are also considered, as they are needed later.

Theorem 1 (generalized from [15]) *Let $P_{XZ'}$ be an arbitrary probability distribution and let z' be a particular value of Z' observed by Eve. If Eve's Rényi entropy $H_2(X|Z' = z')$ about X is known to be at least c and Claude and Dominique choose $K = h_U(X)$ as their secret key, where h_U is chosen at random (uniformly) from a $\epsilon/2^k$-almost universal class of hash functions from $\mathcal{X}$ to $\{0,1\}^k$, where $\mathcal{X}$ is the support set of X, then*

$$H(K|U, Z' = z') \geq (k - \log \epsilon) - 2^{(k-\log \epsilon)-c}/\ln 2.$$

So, we now have a way to wipe out the information of an adversary, Eve, on the final key: Claude and Dominique publicly agree on a hash function from a previously chosen family and then apply it to X, $K = h_U(X)$. With a universal class of hash functions (i.e., $\epsilon = 1$), the number of output bits must be chosen to be less than $c \leq \min_{z'} H_2(X|Z' = z')$, the difference acting as a security margin. Note that, for binary input spaces, there exist families of hash functions for all input sizes [15].

As explained above, Claude and Dominique need to agree on the same random variable X with reconciliation before processing it through the hash function. Let us explain how the cost of reconciliation can be taken into account. To relate this to Theorem 1, the legitimate parties have to estimate $H_2(X|Z' = zm)$ with z (or m) an outcome of Z (or M). Cachin showed [32] that

$$H_2(X|Z' = zm) \geq H_2(X|Z = z) - |M| - 2s - 2$$

with probability $1 - 2^{-s}$. Here, $|M|$ is simply the number of bits revealed during reconciliation, which Claude and Dominique can easily count. So, they need only to estimate $\min_z H_2(X|Z=z)$ and further reduce the number of output bits by the number of revealed bits (plus $2s+2$).

This is important in the sense that the cost of reconciliation can be computed independently of the eavesdropping on the quantum channel. This also justifies the two-step approach of secret-key distillation, namely reconciliation followed by privacy amplification.

The estimation of the Rényi entropy, needed to size the output of the hash functions correctly, is not an easy task. Let us compare it with the Shannon entropy and review one of the difficulties of working with this family of results.

The reason for requiring bounds on the Rényi entropy, instead of the Shannon entropy, is that the Rényi entropy somehow imposes a penalty on non-uniform distributions, as explained in Section 3.4. I will now give an example to show why Rényi entropy, instead of Shannon entropy, is used [15, 32]. By contradiction, let us imagine that a constraint on the Shannon entropy of the eavesdropper's knowledge is imposed. Requiring that $H(X|Z=z) \geq \log|\mathcal{X}| - \epsilon$ (assuming uniform X) may either mean that Eve gets ϵ bits from Z every time (case 1), or that she gets the correct value X with probability $(\epsilon-1)/(\log|\mathcal{X}|-1)$ and a random value in $\mathcal{X} \setminus \{X\}$ otherwise but she is not informed of which event happens (case 2). In case 1, it is legitimate to apply a hash function to spread Eve's uncertainty, whereas in case 2, applying the hash function will not reduce her knowledge in the event that she gets the correct value (i.e., she feeds the correct bits into the hash function). The Rényi entropy, on the other hand, is upper bounded by the Shannon entropy, $H_2(X) \leq H(X)$, with equality only in the case of uniform distribution. In case 2, it is easy to verify that $H_2(X|Z=z) \ll H(X|Z=z)$, so that Rényi entropy correctly addresses this case.

A counter-intuitive property of the Rényi entropy is that $H_2(X|Y) > H_2(X)$ can happen, in contrast to Shannon entropy. In the case of privacy amplification, this may mean that the estimation of the Rényi entropy is a pessimistic approach to what is really possible. Consider the following thought experiment: in addition to the eavesdropped information Z, an oracle gives Eve another variable Z_{extra} correlated to X and Z, such that $H_2(X|ZZ_{\text{extra}}) > H_2(X|Z)$. Eve is free not to use this extra information, but in view of Theorem 1, this actually increases the number of bits one can distill. This is sometimes called *spoiling knowledge* [15, 32], and makes it difficult to estimate the number of bits one can really extract and to optimize the SKD protocol.

6.3.2 Privacy amplification with extractors

Roughly speaking, an extractor is a function that extracts uniformly random bits from a weakly random source using a small number of additional random bits used as a catalyst [139, 145, 172]. Extractors were first defined by Nisan and Zuckerman [139] in the context of theoretical computer science. They have a wide variety of applications in protocol and complexity analysis, and in particular they allow the simulation of randomized algorithms even when only weak sources of randomness are available.

Definition 3 ([139]) *Let $U^{(t)}$ be a random variable with uniform distribution over $\{0,1\}^t$. A function $E : \{0,1\}^l \times \{0,1\}^t \to \{0,1\}^k$ is called a (δ, ϵ)-extractor if for any random variable X with range $\{0,1\}^l$ and min-entropy $H_\infty(X) \geq \delta l$, the variational distance between the distribution of $[U^{(t)}, E(X, U^{(t)})]$ and that of $U^{(t+k)}$ is at most ϵ. (The variational distance between two distributions P_1 and P_2 over the same range is defined as $\sum_x |P_1(x) - P_2(x)|/2$.)*

We assume again that Claude and Dominique performed reconciliation, and thus agreed on a common variables, here denoted X. It is shown by Maurer and Wolf [122] that one can use extractors to perform privacy amplification.

Theorem 2 (based on [122]) *Let δ, Δ_1, $\Delta_2 > 0$ be constants. For a sufficiently large l, let $P_{XZ'}$ be an arbitrary probability distribution with the support of X in $\{0,1\}^l$. Let z' be a particular value of Z' observed by Eve such that Eve's min-entropy $H_\infty(X|Z' = z')$ about X is known to be at least δl. Then there exists a function $E : \{0,1\}^l \times \{0,1\}^t \to \{0,1\}^k$, with $t \leq \Delta_1 l$ and $k \geq (\delta - \Delta_2)l$, such that when Claude and Dominique choose $K = E(X, U)$ as their secret key, where U are randomly and uniformly chosen bits, they obtain*

$$H(K|U, Z' = z') \geq k - 2^{-l^{1/2 - o(1)}}.$$

As for privacy amplification with hash functions, an extractor can spread Eve's uncertainty so that the final key K is virtually secret. There are a few differences, though. First, the eavesdropped information must now be measured in terms of min-entropy instead of order-2 Rényi entropy. Another difference is that the number of random bits to input in the function (U) can be much lower for extractors than for hash functions. This can be useful in some circumstances, e.g., when using these results for the broadcast channel [122], see Section 6.4.1. Finally, Theorem 2 proves the existence of such extractors with sufficiently large input sizes, but does not guarantee their existence for all parameter sizes.

As a last remark, note that we need to take into account the knowledge of reconciliation information. Similarly to privacy amplification with hash functions, it is shown in [122] that

$$H_\infty(X|Z' = zm) \geq H_\infty(X|Z = z) - |M| - s$$

with probability $1 - 2^{-s}$.

6.4 Authenticated repetitive secret-key distillation

In this section, we consider the case of authenticated SKD from multiple outcomes of discrete random variables X, Y and Z, as illustrated in Fig. 6.3. We will consider one-way and two-way communications separately.

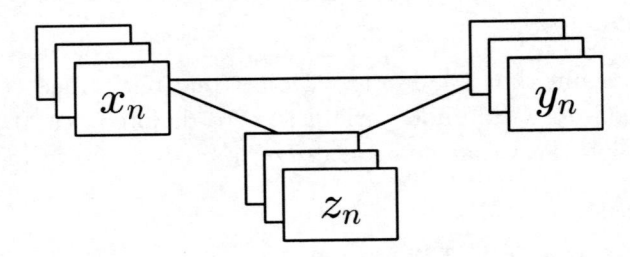

Fig. 6.3. In the repetitive setting, Claude, Dominique and Eve have access to many outcomes of the random variables X, Y and Z.

6.4.1 One-way communications

With one-way communications, we assume that Claude is the one who sends information to Dominique. As a consequence, it must be assumed that the key is a function of Claude's variable, $K = K(X)$.

Formally, the variables Y and Z can be seen as the result of sending X over a broadcast channel characterized by the conditional joint distribution $P_{YZ|X}$. The problem then comes down to determining the number of secret bits that Claude can send to Dominique, which is defined as the *secret capacity* of the corresponding broadcast channel.

The original definition of the secret capacity was given by Wyner [189], who considered the case where Eve can eavesdrop only on the channel output (Dominique's received value) via an additional independent channel.

Csiszár and Körner [48] then generalized Wyner's work by considering that eavesdropping can also occur on the emitter's side.

Definition 4 ([48, 122, 189]) *The (strong) secrecy capacity $\bar{C}_S(P_{YZ|X})$ of a memoryless broadcast channel characterized by the conditional joint distribution $P_{YZ|X}$ is the maximal rate $R \geq 0$ such that $\forall \epsilon > 0$, there exist $l > 0$, a (probabilistic) encoding function $e : \{0,1\}^k \to \mathcal{X}^l$ and a (probabilistic) decoding function $d : \mathcal{Y}^l \to \{0,1\}^k$, with $k = \lfloor (R-\epsilon)l \rfloor$ and $\mathcal{X}$ (or $\mathcal{Y}$) the support of X (or Y), such that for K uniformly distributed over $\{0,1\}^k$, for $X_{1...l} = e(K)$ and $K' = d(Y_{1...l})$, we have $\Pr[K \neq K'] \leq \epsilon$ and $H(K|Z_{1...l}) > k - \epsilon$.*

Note that the original definition, denoted $C_S(P_{YZ|X})$, has weaker secrecy requirements. However, Maurer and Wolf showed [122] that

$$\bar{C}_S(P_{YZ|X}) = C_S(P_{YZ|X}),$$

so we will drop the bar.

The secret capacity is determined as follows:

Theorem 3 ([48]) $C_S(P_{YZ|X}) = \max_{P_{UX}}[I(U;Y) - I(U;Z)]$, *where* $P_{UXYZ} = P_{UX}P_{YZ|X}$. *In the case that* $I(X;Y) \geq I(X;Z)$ *for all choices of* P_X, *then*

$$C_S(P_{YZ|X}) = \max_{P_X}[I(X;Y) - I(X;Z)].$$

The maximization over the probabilities of X are considered by Csiszár and Körner [48] since they deal with a broadcast channel. In the case that the probability of X is fixed (e.g., imposed by the modulation of the chosen QKD protocol), the maximization must be done over U only, if applicable. In particular, Theorem 3 implies that one can distill $I(X;Y) - I(X;Z)$ secret key bits, using asymptotically large block codes.

The original proof of this theorem is based on random coding arguments. Interestingly, the proof of the strong secrecy capacity is based on extractors instead [122].

6.4.2 Two-way communications

When two-way communications are allowed, Dominique can send some information back to Claude. In this case, one has to go beyond the concept of a broadcast channel, as the final key can depend on both variables X and Y. We now consider the case where Claude and Dominique work with a predefined protocol with messages going in both directions. At the end of the protocol, the two legitimate parties must be able to determine K_C and K_D, which are equal with arbitrarily high probability, and which contain arbitrarily secret bits. For a given joint distribution P_{XYZ}, one formally defines the *secret key rate* as in the next definition.

Definition 5 ([117, 122]) *The (strong) secret key rate of X and Y with respect to Z, denoted by $\bar{S}(X;Y \| Z)$, is the maximal rate $R \geq 0$ such that $\forall \epsilon > 0$ and $\forall l \geq N_0(\epsilon)$ there exist a variable K with range $\mathcal{K}$ and a protocol, using public communications over an insecure but authenticated channel, with l independent instances of the variables given: $X_{1...l}$ to Claude, $Y_{1...l}$ to Dominique and $Z_{1...l}$ to Eve, such that Claude and Dominique can compute keys K_C and K_D respectively, verifying $\Pr[K_C = K \wedge K_D = K] \geq 1 - \epsilon$, $I(K; Z_{1...l}M) \leq \epsilon$ and $H(K) = \log|\mathcal{K}| \geq l(R - \epsilon)$, with M the collection of messages sent by Claude and Dominique over the insecure channel.*

Note that the original definition given by Maurer [117], denoted $S(X;Y \| Z)$, has weaker secrecy requirements. However, Maurer and Wolf showed that $\bar{S}(X;Y \| Z) = S(X;Y \| Z)$ [122], so we will again drop the bar.

A closed form of $S(X;Y \| Z)$ is unknown, and upper and lower bounds are stated in Theorems 4 and 5.

Theorem 4 ([117]) *The secret key rate $S(X;Y \| Z)$ is lower-bounded as*

$$I(X;Y) - \min[I(X;Z), I(Y;Z)] \leq S(X;Y \| Z).$$

It is interesting to note that the lower bound in Theorem 4 is based on Theorem 1. For a high number of instances of X, Y and Z, the asymptotic equipartition property [46] says that most of the time, $X^{(l)}$, $Y^{(l)}$ and $Z^{(l)}$ belong to the set of typical sequences, which are uniformly distributed. Conditionally to $X^{(l)}Y^{(l)}Z^{(l)}$ being typical (whose probability can be arbitrarily high), the Rényi entropy matches the Shannon entropy. This allows us to say that the number of secret bits achievable is at least the number of produced bits by Claude $H(X)$, decreased by the number of bits revealed for Dominique to know X, namely $H(X|Y)$, and decreased by the number of bits known to Eve $I(X;Z)$, hence

$$H(X) - H(X|Y) - I(X;Z) = I(X;Y) - I(X;Z).$$

Of course, the same argument is valid if the key is based on Dominique's bits, hence $I(X;Y) - I(Y;Z)$ is also possible.

The first upper bound of $S(X;Y \| Z)$ was given by Maurer [117], namely

$$S(X;Y \| Z) \leq \min[I(X;Y), I(X;Y|Z)].$$

This can be explained intuitively by the fact that the number of secret bits Claude and Dominique can distill cannot be larger than the number of bits they share a priori or than the number of bits they share outside Eve's knowledge of Z.

Then, Maurer and Wolf developed the concept of intrinsic mutual information [119, 120] to improve the upper bound. As a thought experiment, the upper bound above is also valid for any variable $\bar{Z}$ obtained by sending Z through some arbitrary channel. The *intrinsic mutual information* [119, 120] between X and Y given Z is

$$I(X;Y \downarrow Z) = \inf_{P_{\bar{Z}|Z}} I(X;Y|\bar{Z}),$$

with $P_{XYZ\bar{Z}} = P_{XYZ}P_{\bar{Z}|Z}$. Then, we obtain the upper bound $S(X;Y \parallel Z) \leq I(X;Y \downarrow Z)$.

Finally, the upper bound of $S(X;Y \parallel Z)$ was extended by Renner and Wolf [147] by considering that if Eve knows an additional variable Z_{extra}, we have $S(X;Y \parallel Z) \leq S(X;Y \parallel ZZ_{\text{extra}}) + H(Z_{\text{extra}})$. The secret key rate is then minimized using all such possible additional variables.

The *reduced intrinsic mutual information* between X and Y given Z is

$$I(X;Y \downarrow\downarrow Z) = \inf_{P_{Z_{\text{extra}}|XYZ}} (I(X;Y \downarrow ZZ_{\text{extra}}) + H(Z_{\text{extra}})).$$

Theorem 5 ([147]) *The secret key rate $S(X;Y \parallel Z)$ is upper-bounded as*

$$S(X;Y \parallel Z) \leq I(X;Y \downarrow\downarrow Z).$$

At this time of writing, bounds tighter than Theorem 5 are not known.

In practice, the results of secret-key distillation using two-way communications are not easy to exploit. Using the two-step approach, namely reconciliation followed by privacy amplification, one is bound to the limit $I(X;Y) - \min[I(X;Z), I(Y;Z)]$. Since practical coding techniques do not meet Shannon entropies, the practical result is more pessimistic, namely

$$S_{\text{practical}} < I(X;Y) - \min[I(X;Z), I(Y;Z)].$$

In principle, two-way communications would allow one to beat this formula. For instance, consider the case of three binary random variables X, Y and Z, with $\mathcal{X} = \mathcal{Y} = \mathcal{Y} = \{0,1\}$. Each of these variables is balanced, that is, $P_X(0) = P_X(1) = P_Y(0) = P_Y(1) = P_Z(0) = P_Z(1) = 1/2$. They are pairwise correlated, but not perfectly: $\Pr[X = Y] = 1 - \epsilon_{XY}$, $\Pr[Y = Z] = 1 - \epsilon_{YZ}$ and $\Pr[Z = X] = 1 - \epsilon_{ZX}$, for some real values $0 < \epsilon_{XY}, \epsilon_{YZ}, \epsilon_{ZX} \leq 1/2$. Maurer showed [117] that even if $\epsilon_{XY} > \epsilon_{ZX}$ and

$\epsilon_{XY} > \epsilon_{YZ}$, it is possible to distill a secret key using two-way communications. Or stated otherwise, a non-zero secret key rate is possible even if $I(X;Z) > I(X;Y)$ and $I(Y;Z) > I(X;Y)$. However, this is only one case and unfortunately no universal construction is known.

6.5 Unauthenticated secret-key distillation

So far, we have considered the case where the public channel is authenticated outside the distillation protocol. In this section, we discuss and present results about the case where the authentication is not assumed and security against active enemies must be ensured as a part of the distillation protocol itself.

The idea of explicitly dealing with an active enemy, in the scope of secret-key distillation, was first investigated by Maurer [118]. His study was then generalized with Wolf, and their joint results were published as three papers: [123], which deals with general results and repetitive unauthenticated SKD, [124], which presents a novel way to determine whether two distributions are simulatable (see Definition 7), and [125], which presents results about privacy amplification with universal families of hash functions and extractors for one-shot unauthenticated SKD.

We now focus on the repetitive results of [123]. The definition of the secret key rate is adapted to the unauthenticated case as follows.

Definition 6 ([123]) *The robust secret key rate of X and Y with respect to Z, denoted by $S^*(X;Y \parallel Z)$, follows the same definition as Definition 5 with the modification that the channel between Claude and Dominique is not authenticated and that, for any $\delta > 0$, with probability $1 - \delta$ either both parties reject the protocol (if Eve modifies the messages between Claude and Dominique) or the protocol is successful.*

The main result of [123] states that if Eve can somehow mimic Claude (or Dominique), she will be indistinguishable from Claude (or Dominique) to the other party. More formally, whether Eve can mimic either party is determined by the joint probability distribution P_{XYZ}: given Z, she may have the possibility to derive new random variables so as to simulate X or Y. This condition is expressed in Definition 7.

Definition 7 ([118]) *The random variable X is simulatable by Z with respect to Y, denoted $\text{sim}_Y(Z \to X)$, if there exists a conditional distribution $P_{\bar{X}|Z}$ such that $P_{\bar{X}Y} = P_{XY}$.*

6.5 Unauthenticated secret-key distillation

Theorem 6 ([123]) *The robust secret key rate $S^*(X;Y \parallel Z)$ is equal to the secret key rate $S^*(X;Y \parallel Z) = S(X;Y \parallel Z)$, unless either $\mathrm{sim}_Y(Z \to X)$ or $\mathrm{sim}_X(Z \to Y)$, in which cases $S^*(X;Y \parallel Z) = 0$.*

Theorem 6 states that if Eve can simulate either X or Y, distillation cannot work: Claude (or Dominique) may not be able to distinguish Dominique (or Claude) from Eve.

Theorem 6 also states that if Eve cannot simulate either party, SKD works and, given a good enough protocol, the secret key rate does not suffer from the fact that active enemies should also be taken into account. But does it mean that we do not need a secret key to start with? Do we have to rule out what was said in Section 5.1.1? As we will discuss now, the answer is negative: in the scope of QKD, Claude and Dominique still need a (short) secret key to start with.

The price to pay for being able to distill a secret key without a bootstrap secret key is the knowledge of the joint probability distribution P_{XYZ}. From Claude's (or Dominique's) point of view, being able to know what are the statistical differences between Y (or X) and Z is essential and is an assumption that one must keep in mind. Or stated otherwise, the knowledge of P_{XYZ} is to some extent equivalent to Claude and Dominique sharing a bootstrap secret key.

In QKD, the joint probability distribution P_{XYZ} is not known beforehand, but must be estimated. During this estimation, Claude and Dominique exchange a subset of their variables X and Y. This exchange does not need to be secret, as the exchanged key elements are sacrificed and will not be part of the final key, but it needs to be authenticated. In the absence of authentication of this part of the protocol, Claude (or Dominique) cannot tell Dominique (or Claude) and Eve apart, which means that an active enemy could mount a (wo)man-in-the-middle attack.

As a last comment, we must note that the knowledge of the mutual information quantities $I(X;Y)$, $I(X;Z)$ and $I(Y;Z)$ is not always sufficient to determine the simulatability conditions. In particular, given P_{XY}, if $I(X;Y) \leq I(X;Z)$ (or if $I(X;Y) \leq I(Y;Z)$) there always exists a random variable Z, which satisfies the prescribed mutual information quantities, such that $\mathrm{sim}_Y(Z \to X)$ (or $\mathrm{sim}_X(Z \to Y)$).

6.6 Secret-key distillation with continuous variables

In the above sections, Claude's and Dominique's variables X and Y are assumed to be discrete. We now discuss how secret-key distillation can work when X and Y are continuous.

6.6.1 Discrete or continuous components?

An important point is to specify which parts of the distillation protocol need to be discrete or continuous. In particular, we need to address the following questions: "Does the final key need to be continuous as well? And what about the reconciliation messages?" Actually, I will show that both the final key and the reconciliation messages should be discrete, even if the key elements are continuous.

First, a continuous secret key simply would not make much sense in practice. It would need to be used along with a continuous version of the one-time pad, which is possible [159], but which would be difficult to make noise-resistant. It is rather inconvenient to deal with error variances on real numbers, as any resource-limited processing can only spread errors further.

Furthermore, a practically useful continuous one-time pad does not exist. Other continuous encryption schemes exist, but their security is not sufficient in the context of QKD. For instance, there exist sound (e.g., voice) scrambling techniques that process analog signals (e.g., [6, 93]), but they do not achieve perfect secrecy. Another example is cryptography based on synchronized chaos [49]. There, the encrypted signal is simply added to some chaotic noise. The security lies in the fact that the noise has a larger amplitude than the signal for the entire spectrum. Again, this is weaker than perfect secrecy.

Second, the reconciliation messages can be either continuous or discrete. Unless the public authenticated classical channel has infinite capacity, exchanged reconciliation messages are either discrete or noisy continuous values. The latter case introduces additional uncertainties into the protocol, which go against our purposes. For instance, a privacy amplification protocol (e.g., based on hash functions) is aimed at spreading the uncertainty over the entire key, independently of the origin of the uncertainty. Thus, noise in exchanged messages will be spread just like Eve's uncertainty is spread.

Finally, a noisy continuous reconciliation message would benefit less efficiently from the authentication feature of the classical channel. An authentication protocol will have a very hard time recognizing noise due to an active adversary against noise intrinsically present in the messages.

Hence, a discrete final key and discrete reconciliation messages are clearly preferred.

6.6.2 The class of discrete-key distillation from continuous variables

We now investigate the class C of protocols that distill a *discrete* secret key from *continuous* variables using *discrete* messages.

If we impose the conversion of the variables X and Y into discrete variables, say $X' = T_C(X)$ and $Y' = T_D(Y)$, before their processing, we obtain another class C' of protocols, which is clearly contained in the former class, $C' \subseteq C$. We can show that these two classes are actually equal. This means that there is no penalty on the efficiency of the distillation by requesting that X and Y are converted into X' and Y' prior to reconciliation and privacy amplification.

The process of distillation can be summarized as functions $k = f_C(x, m)$ and $k = f_D(y, m)$ to produce the key k, where m indicates the exchanged messages. As both k and m are to be taken in some countable set, these two functions each define a countable family of subsets of values that give the same result: $S_{km} = \{x : f_C(x, m) = k\}$ and $S'_{km} = \{y : f_D(y, m) = k\}$. The identification of the subset in which x (or y) lies is the only data of interest – and can be expressed using discrete variables – whereas the value within that subset does not affect the result and can merely be considered as noise. Obviously, there exist discretization functions T_C and T_D and a discrete protocol represented by the functions f'_C and f'_D such that $f'_C(T_C(x), m) = f_C(x, m)$ and $f'_D(T_D(y), m) = f_D(y, m)$.

Also, the discrete conversion does not put a fundamental limit on the resulting efficiency. It is possible to bring $I(T_C(X); T_D(Y))$ as close as desired to $I(X; Y)$. On the other hand, no fully continuous protocol (i.e., including continuous messages or resulting in a continuous key) can expect Claude and Dominique to share more secret information than they initially share, $I(X; Y)$. Hence, $C' = C$.

Yet, if the conversion to discrete variables prior to secret-key distillation is as efficient as a continuous distillation followed by a conversion to discrete variables, why would one favor the first approach? The reason for this is that processing real variables is prone to errors due to the finite precision of any realistic computer. Processing the real values means also processing irrelevant pieces of information (i.e., those that do not influence the functions f_C and f_D above), thereby wasting resources at the cost of a decrease of precision for the relevant pieces of information.

For all the reasons stated above, an optimal secret-key distillation protocol should consist of converting Claude's and Dominique's continuous variables into discrete ones and exchanging discrete information between the two communicating parties so as to distill a discrete secret key.

6.7 Conclusion

In this chapter, I reviewed the possible techniques to perform secret-key distillation and their related theoretical results. I also addressed the special case of continuous variables.

Efficient secret-key distillation is essential to a QKD-based cryptosystem, in the sense that Claude and Dominique need to deliver a close-to-optimum amount of common secret key bits while controlling the eavesdropper's knowledge on them.

In the following chapters, I will develop further a subset of the techniques discussed in this chapter. In particular, I will explicitly split SKD into reconciliation and privacy amplification. Following a top-down approach, I will first consider privacy amplification using universal families of hash functions, as they work best in practice. Then, I will discuss reconciliation, so as to show how the two parties can actually obtain equal intermediate keys.

7
Privacy amplification using universal families of hash functions

In this chapter, I will discuss some important aspects of universal families of hash functions. I will not remain completely general, however, as we are only interested in universal families of hash functions for the purpose of privacy amplification of QKD-produced bits. In the first section, I explain my motivations, detailing the requirements for families of hash functions in the scope of privacy amplification. I then give some definitions of families and show how they fit our needs. Finally, I discuss their implementation.

Defined in Section 6.3.1, the essential property of an $\epsilon/|\mathcal{B}|$-almost universal family of hash function is recalled in Fig. 7.1.

7.1 Requirements

For the purpose of privacy amplification, families of hash functions should meet some important requirements. They are listed below:

- The family should be universal ($\epsilon = 1$) or very close to it ($\epsilon \approx 1$).
- The number of bits necessary to represent a particular hash function within its family should be reasonably low.
- The family should have large input and large output sizes.
- The evaluation of a hash function within the family should be efficient.

The first requirement directly affects the quality of the produced secret key. The closer to universality, the better the secrecy of the resulting key – see Section 6.3.1.

The second requirement results from the fact that the hash function will be chosen randomly within its family and such a choice has to be transmitted between Claude and Dominique. It is not critical, however, because the choice of the hash function need not be secret. A number of bits proportional to the input size is acceptable.

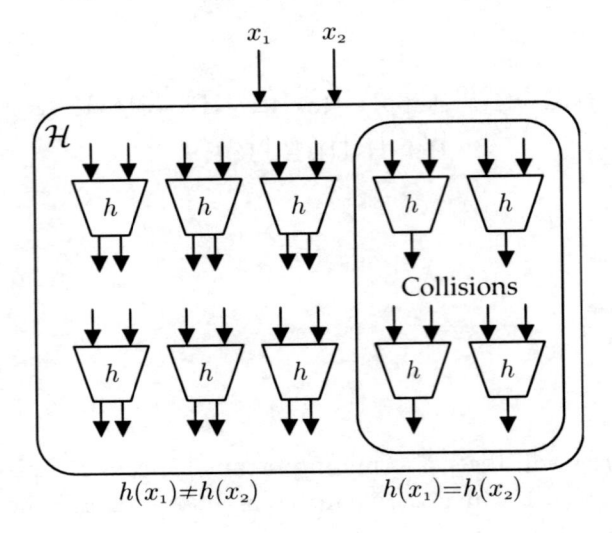

Fig. 7.1. For privacy amplification to work best, a family of hash functions should be such that for any $x_1 \neq x_2$, the set of functions that give the same output $h(x_1) = h(x_2)$ (i.e., a collision) should be as small as possible. To be an $\epsilon/|\mathcal{B}|$-almost universal family of hash function, it may not give more than $\epsilon|\mathcal{H}|/|\mathcal{B}|$ collisions for any $x_1 \neq x_2$, with $|\mathcal{H}|$ the number of functions in the family and $|\mathcal{B}|$ the size of the output set.

Let us explain the need for large input and output sizes. In QKD, the estimation of the number of bits to remove by privacy amplification is statistically determined by Claude and Dominique, who compare some of their samples. Such an estimation must use test samples that are randomly and uniformly spread over a block of a QKD run, otherwise allowing a time-dependent eavesdropping to be incorrectly estimated. Increasing the number of test samples improves the statistical estimation but, on the other hand, it also decreases the number of samples available for the key. The ideal situation would be to have a huge block size, out of which a large number of test samples can be extracted, but whose proportion in the block size is very small. As a trade-off, we look for families with as large as possible input size, with a proportionally large output size.

For instance, the hash functions used for privacy amplification in [77] could process 110 503 bits as input and could produce any smaller number of output bits. These bits came from a block of size of about 55 200 or 36 800 quantum samples, depending on whether we would extract two or three bits out of each sample. The optical implementation gave bursts of 60 000 pulses.

7.1 Requirements

Finally, the efficiency of hash function evaluation is of high practical importance. In a real-time application of QKD, the secret key distillation should not take too much time. The evaluation of the hash function is on the critical path, along with the other steps of the secret-key distillation.

The input and output size requirements of families used for authentication are quite different from those for privacy amplification: for authentication, the output size is quite small and does not grow – or grows slowly – with the input size, see Section 5.1.1. For privacy amplification, however, the output size is proportional to the input size.

In the literature, one can find many families that were constructed with authentication in mind – see for instance [169] and the references therein, or the UMAC construction [24]. Unfortunately, all these families have small output sizes.

7.1.1 Combining and extending families

It is natural to ask whether we can create new families of hash functions by combining or extending existing ones. To some extent, the answer is positive; however, the results in this direction are not very satisfying as they do not meet our requirements.

First, let us consider dividing the input x into two parts $x = x_1 x_2$ (where the invisible composition law is the concatenation) and process these two inputs with two different hash functions and then concatenate the two results $h(x) = h_1(x_1)h_2(x_2)$. Unfortunately, this procedure is not very effective. Roughly speaking, each hash function mixes the bits of its input, but nothing mixes the bits of the two inputs together. In a sense, all the input bits should be mixed together to give as independent as possible output bits. Stinson proved the following result [169]. Let $\mathcal{H} = \{h_i\}$ be an $\epsilon/|\mathcal{B}|$-almost universal family from $\mathcal{A}$ to $\mathcal{B}$, and define $\mathcal{H}^t$ as its Cartesian product $\mathcal{H}^t = \{h_i^t\}$, with $h_i^t(x_1, x_2, \ldots x_t) = (h_i(x_1), h_i(x_2), \ldots h_i(x_t))$. Then, $\mathcal{H}^t$ is $\epsilon'/|\mathcal{B}|^t$-almost universal with $\epsilon' = \epsilon |\mathcal{B}|^{t-1}$. Even if $\mathcal{H}$ is universal ($\epsilon = 1$), the resulting family clearly steps away from universality ($\epsilon' > 1$).

Second, let us consider two families of hash functions: $\mathcal{H}_1$ from $\mathcal{A}_1$ to $\mathcal{B}_1$, which is $\epsilon_1/|\mathcal{B}_1|$-almost universal, and $\mathcal{H}_2$ from $\mathcal{A}_2 = \mathcal{B}_1$ to $\mathcal{B}_2$, which is $\epsilon_2/|\mathcal{B}_2|$-almost universal. Then, the set of compositions $h(x) = h_2(h_1(x))$, $\mathcal{H} = \{h = h_2 \circ h_1 : h_i \in \mathcal{H}_i\}$, is $\epsilon/|\mathcal{B}_2|$-almost universal with $\epsilon = \epsilon_1 |\mathcal{B}_2|/|\mathcal{B}_1| + \epsilon_2$. Unfortunately, even if the two families are universal ($\epsilon_{1,2} = 1$), the resulting one may not be so ($\epsilon_1 |\mathcal{B}_2|/|\mathcal{B}_1| + \epsilon_2 > 1$).

Third, the construction presented in Section 5.1.1 iterates a smaller hash function from $2s$ bits to s bits to construct a larger one. It helps in achieving large input sizes, but the resulting output size is still limited by that of the underlying hash function.

Finally, to increase the output size, it may be interesting to use hash functions with small output sizes to build larger output sizes. This is indeed possible. Let $\mathcal{H}_1$ be a universal family of hash functions with input $\mathcal{A}$ and output $\mathcal{B}_1$. Let $\mathcal{H} = \{(h_1, h_2) \mid h_1, h_2 \in \mathcal{H}_1\}$ be the set of hash functions whose output is the concatenation of the output of two independently chosen hash functions. Collisions in $\mathcal{H}$ occur only when both component functions h_1 and h_2 cause a collision. The number of collisions in $\mathcal{H}$ is thus the product of the number of collisions in each composing family,

$$|\{(h_1(x_1), h_2(x_1)) = (h_1(x_2), h_2(x_2))\}| \leq |\mathcal{H}_1|^2/|\mathcal{B}_1|^2 = |\mathcal{H}|/|\mathcal{B}|.$$

The resulting family is therefore universal if the composing family is universal. This construction can be iterated and it thus gives us a way to arbitrarily increase the output size.

The drawback of this method, however, is that it greatly increases the complexity of the evaluation. Let $k_1 = \log |\mathcal{B}_1|$ be the number of output bits of $\mathcal{H}_1$ and let this number be fixed. To output $\log |\mathcal{B}|$ bits, the composed hash function must make $\log |\mathcal{B}|/k_1$ evaluations of functions of $\mathcal{H}_1$. This number is proportional to the number of output bits, which is proportional to the number of input bits. Assuming that the complexity of evaluation of a $h \in \mathcal{H}_1$ is at least proportional to the number of input bits $l = \log |\mathcal{A}|$, the resulting hash function becomes quadratic in the number of input/output bits. (A hash function should depend on all input bits, hence the assumption of a linear or higher complexity.) This construction is thus general but unfortunately not computationally efficient.

To conclude this section, we have to say that the output size, more than the input size, is critical. Trying to increase the output (and input) size artificially, as in the first construction, causes the family to step away from universality. Efficient hash functions with large input size exist in the scope of authentication, but they usually have small output sizes. One can increase the output size of such functions by way of the last construction, but it is costly. So, for the purpose of privacy amplification, we need efficient families of hash functions that have *intrinsically large input and output sizes*.

7.2 Universal families suitable for privacy amplification

We now list a few universal families of hash functions and discuss their suitability for privacy amplification.

7.2.1 Binary matrices

Definition 8 Let $\mathcal{A} = \mathrm{GF}(2)^l$ and $\mathcal{B} = \mathrm{GF}(2)^k$. For M, a $k \times l$ binary matrix, let $h_M(x) = Mx$ be the product of M with the column vector x. Then, $\mathcal{H}_3 = \{h_M : M \in \mathrm{GF}(2)^{k \times l}\}$ is universal [38].

In this family, the identification of the hash function, namely the matrix M, requires kl bits, which is unfortunately not acceptable for our application. E.g., with $l \approx 10^5$, this would require the order of 10^{10} bits to transmit. Furthermore, its evaluation has a quadratic cost, $O(kl) = O(l^2)$ since $k = O(l)$.

Fortunately, one can restrict the size of the family above by requiring that the matrix M be a Toeplitz matrix, that is, if $M_{i,j} = M_{i+\delta, j+\delta}$ for any $i, j, \delta \in \mathbf{N}$ such that $1 \leq i, i+\delta \leq k$ and $1 \leq j, j+\delta \leq l$. The resulting family $\mathcal{H}_{3,\text{Toeplitz}}$ is still universal [104, 114]. The advantage is that an $k \times l$ Toeplitz matrix requires only $k + l - 1$ bits to transmit: it is entirely determined by its first row and its first column.

7.2.2 Modular arithmetic

Definition 9 Let $\mathcal{A} = \{0, 1, \ldots \alpha - 1\}$ and $\mathcal{B} = \{0, 1, \ldots \beta - 1\}$. Let p be a prime number with $p \geq \alpha$ and $g_{c,d}(x) = (cx + d) \bmod p$. Let $f(x) : \mathbf{Z}_p \to \mathcal{B}$ be any function such that $|\{x \in \mathbf{Z}_p : f(x) = y\}| \leq \lceil p/\beta \rceil$, $\forall y \in \mathcal{B}$. Then the composition of the two functions $h_{c,d}(x) = f(g_{c,d}(x))$ forms the following family: $\mathcal{H}_1 = \{h_{c,d} : c, d \in \mathbf{Z}_p, c \neq 0\}$. Carter and Wegman showed that $\mathcal{H}_1$ is universal [38].

This family allows a compact identification of a hash function. Its only drawback is the use of a non-binary field. For our application, we would prefer to be able to process bits. Is it possible to use arithmetic modulo with a power of two? The family $\mathcal{H}_1$ is not directly usable for a prime power. Instead, let us create a new family in this direction.

Let α and β, $\beta < \alpha$, two strictly positive integers. Let $h_{c,d}(x) = \lfloor (cx + d \bmod 2^\alpha)/2^{\alpha - \beta} \rfloor$, the β most significant bits of the affine function $cx + d$ calculated in the ring of integers modulo 2^α.

Theorem 7 The following family of hash function is universal:
$$\mathcal{H}_{\alpha, \beta} = \{h_{c,d} : c, d \in \mathbf{Z}_{2^\alpha}, \gcd(c, 2) = 1\}.$$

Proof

The size of the input is $|\mathcal{A}| = 2^\alpha$, of the output is $|\mathcal{B}| = 2^\beta$ and of the family is $|\mathcal{H}_{\alpha,\beta}| = 2^{2\alpha-1}$. To prove that this family is universal, one must show that the number of functions in $\mathcal{H}_{\alpha,\beta}$ such that $h_{c,d}(x_1) = h_{c,d}(x_2)$, for any fixed x_1 and x_2, $x_1 \neq x_2$, is upper bounded by $|\mathcal{H}_{\alpha,\beta}|/|\mathcal{B}| = 2^{2\alpha-\beta-1}$.

Let $h_{c,d}(x_1) = h_{c,d}(x_2) = t$, so that $cx_1 + d = 2^{\alpha-\beta}t + u_1$ and $cx_2 + d = 2^{\alpha-\beta}t + u_2$, with $u_1, u_2 \in \mathbf{Z}_{2^{\alpha-\beta}}$. All operations are done modulo 2^α unless stated otherwise. First, let us fix t and u_2. By subtraction, c must verify the equation $c(x_1 - x_2) = u_1 - u_2$. Let $\gcd(x_1 - x_2, 2^\alpha) = 2^\delta$. Since u_2 is fixed, $c(x_1 - x_2)$ must be found in the range $\{-u_2, \ldots, -u_2 + 2^{\alpha-\beta} - 1\}$. Depending on δ, two cases can happen: $\delta < \alpha - \beta$ or $\alpha - \beta \leq \delta < \alpha$. (The case $\delta = \alpha$ is not possible since $x_1 \neq x_2$.) If $\delta < \alpha - \beta$, there are 2^δ possible solutions for c whenever $2^\delta \mid u_1 - u_2$; however, since c must be odd, we must have $2^{\delta+1} \nmid u_1 - u_2$. If $\alpha - \beta \leq \delta < \alpha$, the only possible case for a solution is when $u_1 = u_2$ since $|u_1 - u_2| < 2^{\alpha-\beta}$; but then no odd solution is possible. So, let us concentrate on the case $\delta < \alpha - \beta$. The set $\{-u_2, \ldots, -u_2 + 2^{\alpha-\beta} - 1\}$ contains $2^{\alpha-\beta-\delta-1}$ values such that 2^δ odd solutions can be found. So, for fixed t and u_2, there are up to $2^{\alpha-\beta-1}$ possible values of c. For any given c, the 2^α possible values of (t, u_2) each determine a d. Consequently, there are no more than $2^{2\alpha-\beta-1}$ hash functions in $\mathcal{H}_{\alpha,\beta}$ that give the same output for two different inputs. $\square$

Let us mention a few facts regarding the form of this family. First, the function must be affine for it to be universal; if one takes the subset of linear functions (i.e., where $d = 0$), the resulting family is not universal. The same is true for the requirement that c is odd; allowing c to be even breaks the universality. Finally, taking the least significant bits (instead of the most significant ones) results in a non-universal family.

The identification of a member of the family requires $2l - 1$ bits.

7.2.3 Multiplication in finite fields

Definition 10 *Let $\mathcal{A} = \mathrm{GF}(2^l)$ and $\mathcal{B} = \{0,1\}^k$. Let $h_c(x)$ be defined as the first k bits of the product cx in a polynomial representation of $\mathrm{GF}(2^l)$. The set*

$$\mathcal{H}_{\mathrm{GF}(2^l) \to \{0,1\}^k} = \{h_c : c \in \mathrm{GF}(2^l)\}$$

is a universal family of hash functions [182]. (Note that the location of the k extracted bits does not matter for the family to be universal.)

This family is bit-oriented and requires only l bits to identify a particular function (i.e., the value of c).

Note that the family above does not have to be confused with that of Example 1 in Section 5.1.1. Here, only linear functions are taken, whereas Example 1 requires affine functions. Furthermore, the family in Example 1 is strongly universal, whereas the family in Definition 10 is only universal.

This family will be described in more detail below.

7.3 Implementation aspects of hash functions

In this section, we briefly discuss the implementation of the hash functions mentioned above and then concentrate on the multiplication in a binary field as required by $\mathcal{H}_{\mathrm{GF}(2^l)\to\{0,1\}^k}$.

The $\mathcal{H}_3$ family requires a quadratic time evaluation since all the possible $k \times l$ matrices belong to the family (assuming that k is proportional to l). This is too slow for large input and output sizes.

The subset $\mathcal{H}_{3,\mathrm{Toeplitz}}$ can in fact be seen as a convolution [114] and so can be implemented using Fourier or Fourier-like transforms. Actually, much of what will be said in Section 7.3.1 can in fact be applied to $\mathcal{H}_{3,\mathrm{Toeplitz}}$.

The modular reduction in $\mathcal{H}_{\alpha,\beta}$ is particularly easy to do in a binary representation, as it only requires that we discard the most significant bits. Using the algorithm of Schönhage and Strassen [154], the multiplication of two integers of size l can be performed asymptotically in $O(l \log l \log \log l)$.

7.3.1 Multiplication in a binary field

We now discuss the implementation of the family $\mathcal{H}_{\mathrm{GF}(2^l)\to\{0,1\}^k}$ based on multiplication in binary fields as defined in Definition 10. Among the families presented in Section 7.2, this family was chosen for its lowest number of bits needed to identify a member within the family. Also, it was used for the experimental implementation of QKD with coherent states [77]. In particular, we can describe a possible way to implement this particular operation to process large blocks efficiently.

Multiplying two elements of $\mathrm{GF}(2^l)$ can, of course, be done in quadratic time by using a traditional shift-and-add algorithm. However, this would make the operation quite slow, especially if we need to scale up the block size. We instead describe a simple algorithm that performs a multiplication in $cl \log l$ operations on integers for some small constant c. Actually, this means that $cl \log l$ operations are sufficient as long as numbers of size $\log l$ can be contained in the machine's registers. With 64-bit processors becoming standard, we still have a comfortable margin. In theory, the asymptotic complexity is $O(l \log l \log \log l)$ [155]. The implementation relies on the number-theoretic transform (NTT) [183].

The reduction from a multiplication in $\text{GF}(2^l)$ to the NTT goes as follows:

- First, the multiplication in $\text{GF}(2^l)$ reduces to a multiplication in the ring of binary polynomials $\mathbf{Z}_2[x]$ followed by a reduction modulo $p(x)$, where $p(x)$ is an irreducible polynomial in $\mathbf{Z}_2[x]$ of degree l.
- Second, the multiplication in $\mathbf{Z}_2[x]$ reduces to a multiplication in $\mathbf{Z}[x]$, the ring of polynomials with integer coefficients, followed by a mod 2 reduction of its coefficients.
- Third, the multiplication in $\mathbf{Z}[x]$ reduces to a multiplication in $\mathbf{Z}_m[x]/(x^L-1)$, the ring of polynomials with coefficients in $\mathbf{Z}_m$ and modulo x^L-1 (i.e., the powers of x must be taken mod L, $x^L = x^0 = 1$). This works provided that the polynomials to multiply are limited in their degree and in the value of their coefficients. This is not a problem, since the polynomials to multiply come from $\text{GF}(2^l)$. One can easily check that it works provided that $m > 2l$ and $L > 2l$.
- Finally, the multiplication in $\mathbf{Z}_m[x]/(x^L-1)$ can be done by first transforming the coefficients of both operands using the NTT, multiplying them component-wise, and then transforming back the resulting coefficients using the inverse NTT – see Section 7.3.2.

To implement the first reduction, we need to have an irreducible polynomial of degree l. We are interested in large block sizes, and irreducible polynomials of high degree are not easy to find. In [28], Brent, Larvala and Zimmermann show how to test the reducibility of trinomials (i.e., polynomials of the form $x^l + x^s + 1$) over $\text{GF}(2)$ efficiently. They also propose a list of many irreducible polynomials of degree l with l a Mersenne exponent, that is, such that $2^l - 1$ is a prime. The log files of their implementation can be found in [29], and at this time of writing they propose irreducible polynomials of degrees 127, 521, 607, 1279, 2281, 3217, 4423, 9689, 19 937, 23 209, 44 497, 110 503, 132 049, 756 839, 859 433, 3 021 377 and 6 972 593.

In [77], the chosen field is $\text{GF}(2^{110\,503})$. The reason for choosing a field with degree 110 503 among the possible Mersenne exponents is that the NTT is most efficiently implemented when L is a power of two, $L = 2^\lambda$. Since we must have $L > 2l$, $110\,503 < 131\,072 = 2^{17}$ was closer to the next power of two than $132\,049 < 262\,144 = 2^{18}$ was. To represent this field, the possible irreducible trinomials are $x^{110\,503} + x^{25\,230} + 1$, $x^{110\,503} + x^{53\,719} + 1$, $x^{110\,503} + x^{56\,784} + 1$ and $x^{110\,503} + x^{85\,273} + 1$, where the last two polynomials are reciprocals of the first two [29].

7.3.2 The number-theoretic transform

Let us now detail the multiplication using the NTT. Given two elements of $\mathbf{Z}_m[x]/(x^L - 1)$, say $r(x) = \sum_{i=0}^{L-1} r_i x^i$ and $s(x) = \sum_{i=0}^{L-1} s_i x^i$, their product in this ring is $r(x)s(x) = t(x) = \sum_{i=0}^{L-1} t_i x^i$ with $t_i = \sum_{j=0}^{L-1} r_{i-j} s_j$, where the subscripts must be understood to be modulo L.

Definition 11 *In $\mathbf{Z}_m$, let ω be a L-th root of unity, that is, $\omega^L = 1$ and $\omega^{L'} \neq 1$ for $0 < L' < L$. Given a vector $\mathbf{r} = (r_i)_{i=0...L-1}$ with $r_i \in \mathbf{Z}_m$, define the NTT of $\mathbf{r}$ as $\mathcal{F}\mathbf{r} = \mathbf{R} = (R_j)_{j=0...L-1}$ with $R_j = \sum_{i=0}^{L-1} r_i \omega^{ij}$.*

From now on, we assume that $m = p$ is a prime number. Although NTTs are possible for composite m, they are less efficient [4]. Since ρ-th roots of unity are only available when ρ divides $p - 1$, we must have $p = \nu L + 1$ for some integer ν. For an efficient implementation of NTT (see below), one must have $L = 2^\lambda$ and thus $p = \nu 2^\lambda + 1$.

The NTT can be done in $c_{\text{NTT}} L \log L$ integer operations if L is a power of two, for some constant c_{NTT}. The principle is exactly the same as for the fast Fourier transform (FFT) [44]; one just needs to replace the Lth root of unity in the complex numbers $\omega = e^{i2\pi/L}$ by the Lth root of unity in $\mathbf{Z}_p$ as defined above. For the rest, it works unchanged and the FFT can still rely on the butterfly operation, as $\omega^{L/2} = -1$ in both cases.

The advantage of using NTTs is that convolutions can be performed efficiently. Let the boldface vectors contain the coefficients of the corresponding roman polynomials. Calculating $t(x) = r(x)s(x)$ is equivalent to calculating the convolution $\mathbf{t} = \mathbf{r} * \mathbf{s}$, where $t_j = \sum_{i=0}^{L-1} r_{j-i} s_i$. Using the NTT, $\mathcal{F}(\mathbf{r} * \mathbf{s}) = (\mathcal{F}\mathbf{r}) \cdot (\mathcal{F}\mathbf{s})$, where $\cdot$ indicates the component-wise product, $T_j = R_j S_j$.

To find $\mathbf{t}$ from $\mathbf{T} = \mathcal{F}(\mathbf{t})$, one can use the inverse NTT. It works like the direct one, but with the inverse of the Lth root of unity ω^{-1} instead: $t_i = \sum_{j=0}^{L-1} T_i \omega^{-ij}$.

For instance, let $L = 2^{18} = 262\,144$. $L + 1 = 262\,145 = 5 \times 52\,429$ and $2L + 1 = 524\,289 = 3 \times 174\,763$ are composite, while $3L + 1 = 786\,433 = p$ is prime, thus $\nu = 3$. In $\mathbf{Z}_{786\,433}$, the smallest generator is $g = 11$, that is $g^{p-1} = 1$ while no other smaller exponent achieves this property. Then, by setting $\omega = g^\nu = 11^3 = 1331$, we obtain the 262 144th root of unity ($\omega^{262\,144} = 1$). Note that, as expected, $\omega^{131\,072} = -1$, which enables the butterfly operation. Finally, we need to calculate $\omega^{-1} = 104\,582$ for the inverse NTT.

7.3.3 Family based on number-theoretic transforms

As the NTT seems like a powerful tool, let us leave aside the multiplication in a finite field and see whether we can create a universal family of hash function based on the NTT.

To evaluate a function in $\mathcal{H}_{GF(2^l) \to \{0,1\}^k}$, the NTT provides fast multiplications in the ring $\mathbf{Z}_p[x]/(x^L - 1)$. By working in this ring directly, we avoid the reductions described in Section 7.3.1 and access the NTT more directly for the best implementation performances. More specifically, we could process about $L \log p$ input bits at a time instead of only $l < L/2$.

We actually consider two equivalent families of hash functions. These two families are equivalent from the point of view of universality. One has a form that makes it easy to implement, while the other has a nice algebraic interpretation.

Definition 12 *Let $1 \leq \beta \leq L$. For $\mathbf{C}, \mathbf{R} \in \mathbf{Z}_p^L$ and such that $\mathbf{C}$ has no zero element, $C_i \neq 0 \, \forall i = 0 \ldots L-1$, let $h_\mathbf{C}(\mathbf{R}) = (\mathcal{F}^{-1}(\mathbf{C} \cdot \mathbf{R}))_{0 \ldots \beta - 1}$ be the inverse NTT of their component-wise product, taking only the β first elements of the result. Let us define the family $\mathcal{H}_{p,L,\beta} = \{h_\mathbf{C} \,:\, C_i \neq 0 \, \forall i\}$.*

Definition 13 *Let $1 \leq \beta \leq L$. For $c, r \in \mathbf{Z}_p[x]/(x^L - 1)$ and such that c has a multiplicative inverse, let $h'_c(r) = cr \mod x^\beta$ be their product taking off the powers of x of degree β and higher. Let us define the family $\mathcal{H}'_{p,L,\beta} = \{h_c \,:\, c^{-1} \exists\}$.*

Whether a multiplicative inverse exists in $\mathbf{Z}_p[x]/(x^L - 1)$ can easily be determined from its NTT: an element c has an inverse iff the NTT of its coefficients, $\mathbf{C} = \mathcal{F}(\mathbf{c})$, does not have any zero, $C_i \neq 0 \, \forall i = 0 \ldots L-1$. This follows from the property of $\mathbf{Z}_p$ that each multiplication of the NTT coefficients can be inverted iff the multiplicand is non-zero.

Given $t = cr$, we have $\mathbf{t} = \mathcal{F}^{-1}(\mathcal{F}(\mathbf{c}) \cdot \mathcal{F}(\mathbf{r})) = \mathcal{F}^{-1}(\mathbf{C} \cdot \mathbf{R})$ with $\mathbf{C} = \mathcal{F}(\mathbf{c})$ and $\mathbf{R} = \mathcal{F}(\mathbf{r})$. So, describing the hash function in $\mathcal{H}'_{p,L,\beta}$ with $\mathbf{c}$ or with $\mathbf{C}$ is essentially the same. The same argument applies to the input $\mathbf{r}$. In other words, the two families are equivalent up to permutation of the inputs, so they share the same universality properties: $\mathcal{H}_{p,L,\beta}$ is $\epsilon/|\mathcal{B}|$-almost universal iff $\mathcal{H}'_{p,L,\beta}$ is $\epsilon/|\mathcal{B}|$-almost universal, with $|\mathcal{B}| = p^\beta$ the output size.

Theorem 8 *If $p - 1 \geq L$, both $\mathcal{H}_{p,L,\beta}$ and $\mathcal{H}'_{p,L,\beta}$ are $\frac{p^\beta}{(p-1)^\beta}/|\mathcal{B}|$-almost universal.*

Proof

We prove this only for $\mathcal{H}_{p,L,\beta}$, the result being also directly applicable to $\mathcal{H}'_{p,L,\beta}$.

The size of the input is $|\mathcal{A}| = p^L$, of the output is $|\mathcal{B}| = p^\beta$ and of the family is $|\mathcal{H}_{p,L,\beta}| = (p-1)^L$. To prove the theorem, one must show that the number of functions in $\mathcal{H}_{p,L,\beta}$ such that $h_\mathbf{C}(\mathbf{R}^{(1)}) = h_\mathbf{C}(\mathbf{R}^{(2)})$, for any fixed $\mathbf{R}^{(1)}$ and $\mathbf{R}^{(2)}$, $\mathbf{R}^{(1)} \neq \mathbf{R}^{(2)}$, is upper bounded by $(p-1)^{L-\beta}$.

For a given $\mathbf{R}$, the result of the hash function is composed of the β values $t_i = \sum_{j=0}^{L-1} \omega^{-ij} C_j R_j$, $i = 0 \ldots \beta - 1$. Let $\Delta \mathbf{R} = \mathbf{R}^{(1)} - \mathbf{R}^{(2)}$. For the β values to be equal, we must have

$$\sum_{j=0\ldots L-1} \omega^{-ij} C_j \Delta R_j = 0, \; i = 0 \ldots \beta - 1. \tag{7.1}$$

Let us first consider the system in Eq. (7.1) with the ΔR_j as unknowns. The matrix of the system is the $\beta \times L$ matrix $M = (\omega^{-ij} C_j)_{i,j}$. From the properties of the Vandermonde matrix and from the fact that $C_j \neq 0$, M is of rank β. The requirement that $p - 1 \geq L$ ensures that the matrix does not have proportional columns. To solve the homogeneous system in Eq. (7.1), one must thus have at least β different positions for which $\Delta R_j \neq 0$.

Let us then consider the system in Eq. (7.1) with the C_j as unknowns. The matrix of the system is the $\beta \times L$ matrix $M' = (\omega^{-ij} \Delta R_j)_{i,j}$. Again from the properties of the Vandermonde matrix and from the fact that ΔR_j is non-zero for at least β positions, M' is also of rank β. Therefore, the system $M'\mathbf{C} = 0$ has a vector space of dimension β as solution. Out of these solutions, $(p-1)^{L-\beta}$ have non-zero components.

□

Such hash functions can be implemented very efficiently. One simply has to multiply each component of the input with a randomly chosen integer between 1 and $p - 1$, then apply the NTT to the result, and finally take the desired number of samples. Let us compare with the multiplication in a binary field as implemented in Section 7.3.

Let the number of arithmetic operations needed to perform the NTT to be $c_{\text{NTT}} L \log L$. For a function in $\mathcal{H}_{\text{GF}(2^l) \to \{0,1\}^k}$, we have $L \approx 2l$, hence the number of steps needed to calculate the NTT to evaluate such a hash function is $2c_{\text{NTT}} l \log l$, plus some terms proportional to l. We here disregard all the steps needed by the reduction from $\text{GF}(2^l)$ down to $\mathbf{Z}_m[x]/(x^L - 1)$, as they are all proportional to l. The number of input bits is l, so this gives $2c_{\text{NTT}} \log l$ arithmetic operations per processed bit.

The NTT used to evaluate a hash function in $\mathcal{H}_{p,L,\beta}$ also takes $c_{\text{NTT}} L \log L$. Here, we have $L \approx p$ and the NTT takes about $c_{\text{NTT}} L \log L \approx c_{\text{NTT}} L \log p$

steps. However, the NTT here processes an input size of about $L \log p$ bits, hence giving c_{NTT} operations per processed bits. For instance, with $l \approx 2^{17}$, the evaluation of a function in $\mathcal{H}_{p,L,\beta}$ would be 34 times faster than for $\mathcal{H}_{\text{GF}(2^l) \to \{0,1\}^k}$ with the implementation described above.

Besides this significant practical speed-up, this family suffers from two minor drawbacks. The first one is that the family is not universal – only close to universal. However, the difference with universality may be very small if the chosen prime number p is large enough. The second drawback is that the input and the output are not binary. For the input, this is not a problem, since universality cannot diminish if we restrict the input size to a power of two. Hence, we can immediately use a binary input size. For the output, it may not be a good idea to convert each element of $\mathbf{Z}_p$ into $\lceil \log p \rceil$ bits as the result would not be uniform. However, if one accepts variable-size output, rejection may be a way to obtain uniform bits: since p has the form $p = \nu 2^\lambda + 1$, rejection happens only when $t_i = p - 1$.

7.4 Conclusion

In this chapter, we saw various kinds of families of hash functions and how they apply to privacy amplification purposes. I explained why large input and output sizes are required and why it is important to make the evaluation of such hash functions efficient. Finally, I discussed implementation aspects of universal families of hash functions.

For privacy amplification to work correctly, Alice and Bob must have equal inputs. To convert their correlated measurement and modulation strings, X and Y, into equal strings Ψ, one must correct errors – a process which is discussed in the next two chapters.

8
Reconciliation

Reconciliation is the technique needed to ensure that Claude's and Dominique's key elements are equal. Starting from outcomes of the random variables X and Y, they wish to agree on an equal string Ψ.

In this chapter, I will first give some general properties and then overview and introduce several classes of reconciliation techniques.

8.1 Problem description

The goal of the legitimate parties is to distill a secret key, i.e., to end up with a shared binary string that is unknown to Eve. We assume as a convention that Claude's outcomes of X will determine the shared key K. The common string Ψ before privacy amplification can thus be expressed as a function $\Psi(X)$.

Reconciliation consists in exchanging messages over the public classical authenticated channel, collectively denoted M, so that Dominique can recover Ψ from M and the outcomes of Y. If we denote as $x_{1...l}$ a vector of l independent outcomes of X, the string $\Psi(x_{1...l}) \triangleq (\Psi(x_1), \ldots, \Psi(x_l))$ can be compressed to obtain about $lH(\Psi(X))$ common uniform bits.

As explained in Chapter 6, the impact of reconciliation on privacy amplification is a decrease of $|M|$ bits in the key length, where $|M|$ is the number of bits exchanged during reconciliation.

Our goal is thus to maximize $lH(\Psi(X)) - |M|$, or if Ψ is given, to minimize the number $|M|$ of disclosed bits.

8.1.1 Characteristics of reconciliation protocols

Before dealing with the details of reconciliation, let us here review some of its characteristics.

First, reconciliation can be either *one-way* or *interactive*. The first case is obviously only possible with one-way secret-key distillation, as information is sent only from Claude to Dominique. In the latter case, Claude and Dominique exchange information both ways. The difference between one-way and interactive reconciliation is discussed in Section 8.1.2.

Then, the random variables X and Y to reconcile can either be *discrete* or *continuous*. In both cases, the common string Ψ to obtain is required to be discrete, as detailed in Section 6.6. The case of continuous variables is treated in Chapter 9.

In the case of discrete random variables X and Y, they can be either *binary* when $\mathcal{X} = \mathcal{Y} = \{0,1\}$ or *non-binary* otherwise. We distinguish between binary and non-binary reconciliation since binary reconciliation has been studied more often than non-binary. Examples of binary reconciliation protocols can be found in Sections 8.3 to 8.5. I will also show in Chapter 9 how one can build a non-binary reconciliation protocol based on binary reconciliation protocols as primitives.

Finally, a reconciliation protocol can be oriented towards the encoding of *individual symbols* or *blocks*. This distinction is not essential as blocks can be seen as individual symbols of a larger space. However, having these options in mind may help in the following discussion.

8.1.2 Fundamental limits of reconciliation

Claude and Dominique wish to obtain the same string, and we again assume that the target is Claude's key elements, or a function of it, $\Psi(X)$. The criterion to optimize is the number of disclosed bits needed to obtain the same string. Is interactive reconciliation better than one-way reconciliation? In principle, only Dominique needs to get information from Claude, but interactivity helps in quickly narrowing down the errors to correct.

For one-way reconciliation, the protocol is easy to describe. Claude must send Dominique a function of X, namely $\alpha(X)$ so that s/he can recover X knowing Y. This process, called *source coding with side information*, will be described in detail in Section 8.2 – see also Fig. 8.1.

Source coding with side information is a special case of a more general concept called *distributed source coding*, where two correlated sources are compressed independently. Surprisingly, Slepian and Wolf [167] showed that the two sources can, in principle, be compressed at the same rate as if they were compressed together. This is formalized in Theorem 9.

Theorem 9 ([167]) *Let X and Y be random variables, possibly correlated.*

Let there be two independent coders, one with rate R_X to encode X and the other with rate R_Y to encode Y. The achievable rate region for the decoder to be able to decode both X and Y is given by:

$$R_X \geq H(X|Y), \ R_Y \geq H(Y|X), \ R_X + R_Y \geq H(X,Y).$$

The problem of source coding with side information is a special case of distributed source coding with one of the sources being compressed in an invertible way. This is detailed in Corollary 1.

Corollary 1 ([167]) *Let X and Y be as in Theorem 9. Let R_X be the rate to encode X (without the knowledge of Y). Let Y be given to the decoder. The achievable rate region for the decoder to be able to decode X knowing Y is given by:*

$$R_X \geq H(X|Y).$$

In the perspective of one-way reconciliation, the result of Slepian and Wolf implies that Claude needs to send at least $H(\Psi(X)|Y)$ bits of information. Accordingly, we have to assume that the eavesdropper acquires the same number of bits of information on $\Psi(X)$. Can we do better with interactive reconciliation? In principle, we show that an interactive reconciliation protocol also has to reveal at least $H(\Psi(X)|Y)$ bits on $\Psi(X)$.

Theorem 10 *A reconciliation protocol reveals at least $H(\Psi(X)|Y)$ bits about $\Psi(X)$ in both one-way and interactive cases.*

Proof

For simplicity, let us consider the case of a three-step interactive protocol – it is easy to generalize to a higher number of steps. Claude sends message M_1 to Dominique, who responds with message M_2 to Claude, who in turn sends message M_3 to Dominique. For the first message, $Y \to \Psi(X) \to M_1$ is a Markov chain. Intuitively, it means that information on Y contained in M_1 is only given indirectly through $\Psi(X)$. Markovity implies that

$$I(\Psi(X); M_1) \geq I(\Psi(X); M_1|Y).$$

For the second message, M_1 is known to everybody. As a consequence, the variables $\Psi(X)|M_1 \to Y|M_1 \to M_2|M_1$ form a Markov chain, and

$$I(\Psi(X); Y|M_1) \geq I(\Psi(X); Y|M_1 M_2).$$

The same reasoning applies to the third message; that is, the variables

$Y|M_1M_2 \to \Psi(X)|M_1M_2 \to M_3|M_1M_2$ form a Markov chain and thus

$$I(\Psi(X); M_3|M_1M_2) \geq I(\Psi(X); M_3|YM_1M_2).$$

Using the Markovity relative to the first and third messages, the revealed information

$$I(\Psi(X); M_1M_2M_3) = I(\Psi(X); M_1) \\ + I(\Psi(X); M_2|M_1) + I(\Psi(X); M_3|M_1M_2)$$

can be lower bounded as

$$I(\Psi(X); M_1M_2M_3) \geq H(\Psi(X)|Y) - H(\Psi(X)|YM_1) + H(\Psi(X)|M_1) \\ - H(\Psi(X)|M_1M_2) + H(\Psi(X)|YM_1M_2) - H(\Psi(X)|YM_1M_2M_3) \\ = I(\Psi(X); M_1M_2M_3|Y) + I(\Psi(X); Y|M_1) - I(\Psi(X); Y|M_1M_2).$$

Using the Markovity relative to the second message, we obtain

$$I(\Psi(X); M_1M_2M_3) \geq I(\Psi(X); M_1M_2M_3|Y) = H(\Psi(X)|Y),$$

where the last equality follows from the fact that Dominique can reconstruct $\Psi(X)$ given Y and $M_1M_2M_3$, hence $H(\Psi(X)|YM_1M_2M_3) = 0$. □

The quantity $H(\Psi(X)) - |M|/l$ to maximize is thus upper bounded as

$$H(\Psi(X)) - |M|/l \leq I(\Psi(X); Y) \leq I(X; Y).$$

Consequently, we define the *efficiency of reconciliation* as

$$\eta = \frac{H(\Psi(X)) - |M|/l}{I(X; Y)}.$$

The remainder of this chapter is organized as follows. First, I overview the general problem of source coding with side information and mention some constructions. I then review the existing binary interactive reconciliation protocols. Finally, I introduce turbo codes and low-density parity-check codes.

8.2 Source coding with side information

The problem of source coding with side information at the receiver is to encode the random variable X into a code $\alpha(X)$ such that the receiver, who knows a correlated variable Y, can decode $\bar{X} = \beta(\alpha(X), Y)$ to recover X with no error, $\bar{X} = X$, or possibly with a small error, in which case $\Pr[\bar{X} = X] \approx 1$. This is illustrated in Fig. 8.1. For simplicity, we forget the

8.2 Source coding with side information

words "at the receiver" in the sequel, since we will never consider the case of side information "at the sender". The crucial point is that the encoder does not have access to the variable Y. He knows the joint distribution of XY but he does not know the value of the outcome of Y when encoding an outcome of X.

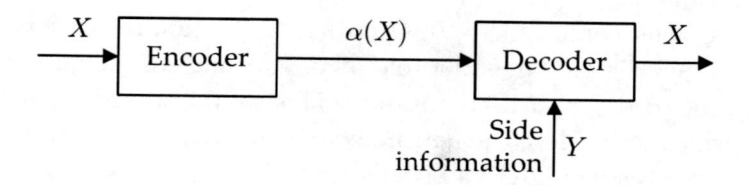

Fig. 8.1. Source coding of X with side information Y known at the decoder.

As proved by Slepian and Wolf [167], source coding with side information can compress the source with a rate not lower than $H(X|Y)$. This is a surprising statement, as it seems that not having access to Y at the encoder does not penalize the encoding rate. If the encoder knew Y, he could encode X and Y jointly using a rate of $H(X,Y)$ bits. The encoding of Y alone would take $H(Y)$ bits, leaving also $H(X,Y) - H(Y) = H(X|Y)$ bits to encode the information contained in X.

In practice, however, the fact that Y is unknown to the encoder makes the encoding more difficult, raising interesting problems.

Used for reconciliation, source coding with side information is typically oriented towards the correction of individual symbols. However, constructions based on syndromes of error-correcting codes are rather oriented towards blocks of symbols – see Section 8.2.4.

8.2.1 Definitions and characteristics

For a given probability distribution $P_{XY}(x,y)$, the symbols $x, x' \in \mathcal{X}$ are said to be *confusable* if there exists $y \in \mathcal{Y}$ such that

$$P_{XY}(x,y) > 0 \text{ and } P_{XY}(x',y) > 0.$$

If the encoder encodes such x and x' with the same codeword, the decoder β may not be able to determine which one was encoded.

A code can be a *zero-error code* or a *near-lossless code*. For a zero-error code, the decoder must always be able to recover X without any error, that is,

$$\Pr[\beta(\alpha(X), Y) = X] = 1.$$

For a near-lossless code, the decoder is allowed a small probability of error. The *probability of confusion* is defined as

$$P_c = \Pr[\beta(\alpha(X), Y) \neq X]. \tag{8.1}$$

For many interesting cases, such as joint Gaussian variables, the joint probability function P_{XY} can be strictly positive for all symbol pairs. All symbols are thus confusable. This means that a non-zero probability of error at the decoder side must be tolerated, allowing some symbols to have identical codewords, even if confusable. This would otherwise make $\alpha(X)$ bijective, which would lose the advantage of side-information. Furthermore, in the particular case of QKD, a bijective encoding scheme would completely disclose X, discarding all the secrecy shared between Claude and Dominique.

The rate of a code with side information is defined as usual and reads $R = \sum_x P_X(x)|\alpha(x)|$. Notice that the rate depends only on the marginal distribution of X.

Alon and Orlitsky [3] define two kinds of zero-error codes: *restricted inputs* (RI) codes and *unrestricted inputs* (UI) codes. By definition, no error is possible with UI and RI codes since no two confusable symbols are assigned to prefix or equal codewords. We add a third kind, namely *near-lossless unrestricted inputs* (NLUI) codes. Oddly enough, UI codes are more restrictive than RI codes.

- A code α is an RI code if $\alpha(x)$ is not a prefix of $\alpha(x')$ whenever x and x' are confusable.
- A code α is a UI code if $\alpha(x) \neq \alpha(x')$ whenever x and x' are confusable and if for all $x, x' \in \mathcal{X}$, $\alpha(x)$ is not a proper prefix of $\alpha(x')$ (even if x and x' are not confusable).
- A code α is an NLUI code if for all $x, x' \in \mathcal{X}$, $\alpha(x)$ is not a proper prefix of $\alpha(x')$.

In general, the codes of consecutive inputs are concatenated to make a binary stream, so we must make sure that the stream produced by α can be properly delimited. Fortunately, the definition of an RI code ensures that α is a prefix-free code when the value of Y is given. For an RI code, an observer not having access to Y may not be able to delimit the codewords in the stream. For UI and NLUI codes, the codewords can be delimited without the knowledge of Y.

For near-lossless coding, we may need to design codes that do assign identical codewords to confusable symbols. This of course leads to decoding errors but allows us to decrease the rate. It is, however, essential to avoid a codeword that is a proper prefix of another codeword, for two confusable symbols. Let x and x' be confusable symbols. If $\alpha(x)$ is a proper prefix of $\alpha(x')$, then $|\alpha(x)| \neq |\alpha(x')|$ and the decoder would not be able to determine the length of the codeword. Consequently, the decoder β would desynchronize, making it impossible to decode the rest of the stream. Using an NLUI code is a possible option for near-lossless coding, as the code can be delimited even without the side information. Confusion can still happen but desynchronization cannot.

The UI and NLUI codes are equivalent to a two-step encoding procedure. First, the set $\mathcal{X}$ of symbols is mapped onto a smaller set by applying a function $\phi : \mathcal{X} \to \mathbf{N} : x \to \phi(x)$. Then, the symbols produced by $\phi(X)$ are encoded using a lossless prefix-free code α_0. The function $\phi(x)$ partitions the symbols $x \in \mathcal{X}$ into sets of equal codewords. The resulting code is zero-error iff for any two confusing symbols x and x' we have $\phi(x) \neq \phi(x')$.

The encoding part α_0 of an (NL)UI code can be Huffman coding or arithmetic coding. For the former, the rate is $R = \sum_c P_{\phi(X)}(c)|\alpha_0(c)|$, where $P_{\phi(X)}(c)$ denotes $\Pr[\phi(X) = c]$, and it verifies $H(\phi(X)) \leq R < H(\phi(X))+1$. For the latter, the rate is very close to the Shannon limit, $R \approx H(\phi(X))$.

Before we give some examples of constructions of codes, I wish to explain the relationship between the zero-error case and some graph properties.

8.2.2 Zero-error codes and graph entropies

Let me define the *confusability graph* and explain its relationship to the problem of zero-error source coding with side information. Consider the random variables X and Y. Let $G = G(X,Y)$ be a graph with vertex set $V(G) = \mathcal{X}$ and edge set $E(G)$. The edge $\{x, x'\}$ belongs to $E(G)$ iff x and x' are confusable.

A *coloring* of G is a map $\phi : V(G) = \mathcal{X} \to \mathbf{N}$ that assigns colors to vertices in such a way that any two adjacent vertices have different colors. The *chromatic number* $\chi(G)$ is the minimum number of colors in any coloring of G. Colorings using exactly $\chi(G)$ colors are referred to as *minimum cardinality colorings*.

A *probabilistic graph* [3, 100] is a pair (G, P) where G is a graph and P a probability distribution over its vertices. For simplicity, we also note as $G(X,Y)$ the probabilistic graph obtained from the confusability graph $G(X,Y)$ and from the probability distribution associated to the random variable X.

The *entropy of a coloring* ϕ of $G(X,Y)$ [3] is the entropy $H(\phi(X))$ of the random variable $\phi(X)$. The *chromatic entropy* $H_\chi(G(X,Y))$ of $G(X,Y)$ is the minimum entropy of any of its coloring,

$$H_\chi(G(X,Y)) = \min\{H(\phi(X)) : \phi \text{ coloring of } G\}.$$

Finally, a *minimum entropy coloring* is a coloring ϕ that achieves the chromatic entropy.

Alon and Orlitsky [3] showed that the minimum rate R_{RI} of an RI code is upper bounded as

$$R_{\text{RI}} \leq H_\chi(G(X,Y)) + 1,$$

and that the lowest asymptotically achievable rate for transmitting multiple instances of X using an RI code is $R_{\text{RI},\infty} = \lim_{d\to\infty} H_\chi(G^{\wedge d}(X,Y))/d$, where $G^{\wedge d}$ is the dth and-power of G. For UI codes, they showed that the minimum rate R_{UI} is lower and upper bounded as

$$H_\chi(G(X,Y)) \leq R_{\text{UI}} \leq H_\chi(G(X,Y)) + 1.$$

Finally, the lowest asymptotically achievable rate for transmitting multiple instances of X using a UI code equals $R_{\text{UI},\infty} = \lim_{d\to\infty} H_\chi(G^{\vee d}(X,Y))/d$, where $G^{\vee d}$ is the dth or-power of G.

The two-step encoding approach of UI codes can be described as first a coloring ϕ of the graph $G(X,Y)$ and then an encoding α_0 of the color information. Optimal UI codes with arithmetic coding, in particular, can achieve rates arbitrarily close to $H_\chi(G(X,Y))$. For NLUI codes, however, ϕ is not necessarily a coloring of $G(X,Y)$, since ϕ is allowed to have $\phi(x) = \phi(x')$ even when x and x' are confusable, that is, when $\{x, x'\} \in E(G)$.

It may be tempting to relate the chromatic entropy to the well-known minimum cardinality coloring problem. For instance, Witsenhausen relate zero-error codes to the chromatic number of the confusability graph [187]. However, it was shown that a minimum cardinality coloring does not necessarily imply the best rate [3]. The chromatic number and the chromatic entropy are actually quite different problems, and some fundamental differences are detailed in [37].

Complexity

It was shown in [37, 195] that finding a minimum entropy coloring is NP-hard.

Theorem 11 ([37]) *Finding a minimum entropy coloring of a graph $G(X,Y)$ is NP-hard, even if X has a uniform distribution, $G(X,Y)$ is planar, and a minimum cardinality coloring of $G(X,Y)$ is given.*

8.2 Source coding with side information

The consequence is that finding optimal zero-error codes is a difficult problem. Unless P = NP, finding a minimum entropy coloring is exponential in $|\mathcal{X}|$. So, approximate solutions are preferred in practice.

8.2.3 Example

As an example, consider the variables X and Y with $\mathcal{X} = \{x_1, x_2, x_3, x_4, x_5\}$ and $\mathcal{Y} = \{y_1, y_2, y_3\}$. The probability distribution $P_{XY}(x, y)$ is defined in Table 8.1.

Table 8.1. *Specification of the joint probability $p_{XY}(x, y)$. An empty entry means that the probability is zero.*

	x_1	x_2	x_3	x_4	x_5
y_1	1/7		1/7		
y_2		1/7	1/7		
y_3			1/7	1/7	1/7

With this probability distribution, the symbols x_1 and x_3 are confusable, since the decoder must be able to distinguish them when $Y = y_1$. Likewise, x_2 and x_3 are confusable, and any pair of $\{x_3, x_4, x_5\}$ also contains confusable symbols. The confusability graph is sketched in Fig. 8.2.

First, an example of RI code is depicted in Fig. 8.2. Notice that the code is not globally prefix-free, as for example the symbols x_2 and x_5 are mapped to codewords 0 and 00, respectively. The code is prefix-free for any pair of confusable symbols, though. The rate of this code is $R_{\text{RI}} = 1 \times 5/7 + 2 \times 2/7 = 9/7$ bits.

Then, an example of UI code is also depicted in Fig. 8.2. Since a UI code must be globally prefix-free, the symbols x_1 and x_2 are now associated to the codeword 00. The rate of this code is higher than that of the RI code since UI codes are more restrictive, $R_{\text{UI}} = 1 \times 3/7 + 2 \times 4/7 = 11/7$ bits.

Finally, Fig. 8.2 also contains an example of NLUI code. The code is globally prefix-free, but the confusable symbols x_4 and x_5 are associated with the same codeword. Assuming that the decoder β takes the arbitrary decision to decode 0 as x_4 when $Y = y_3$, the probability of error is $P_{\text{c}} = P_{XY}(x_5, y_3) = 1/7$. With this error tolerated, the rate is the lowest of all three examples with $R_{\text{NLUI}} = 1$ bit.

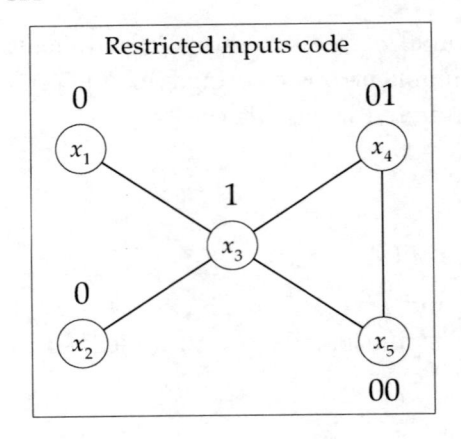

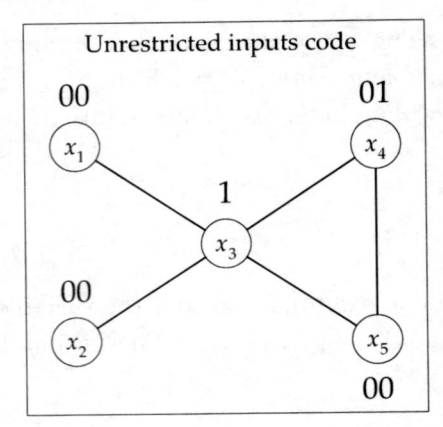

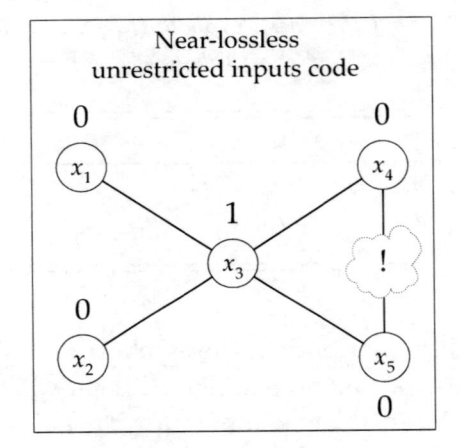

Fig. 8.2. Confusability graph $G(X,Y)$ and examples of codes.

8.2.4 Existing code constructions

I now overview some existing methods to construct zero-error and near-lossless codes.

General constructions

Zhao and Effros [193, 196] propose a construction called multiple access source code (MASC) to produce optimal RI codes, generalizing Huffman-type and arithmetic-type codes. They introduce the notion of partition trees to capture the requirements on which symbols can have equal or prefix codewords. For RI codes, the tree can have several levels; for UI codes, however, the partition tree is flat as codewords may not be proper prefixes of others. Finding the partition tree that leads to the optimal code is difficult.

For a given partition tree, however, finding the optimal code that matches the partition tree is easy.

The same authors propose some fast sub-optimal algorithms based on order-constrained partitions [195]. The idea is to impose an order $\mathcal{O}(\mathcal{X})$ on the symbols of $\mathcal{X}$. The partition can only gather consecutive symbols so as to restrict the number of partitions in the search for an optimal code. Constrained to a given order $\mathcal{O}(\mathcal{X})$, their algorithm can find the optimal partition and associated code in $O(|\mathcal{X}|^4)$. To find a good code, an optimization of $\mathcal{O}(\mathcal{X})$ is performed using either simulated annealing or other heuristic search algorithms. The global performance is $O(|\mathcal{X}|^6)$ [194].

A greedy merger algorithm derived from rate-distortion optimization techniques is proposed in [36]. Its complexity is $O(|\mathcal{X}|^3|\mathcal{Y}|)$.

Further properties of zero-error codes are given by Yan and Berger [190] where necessary or sufficient conditions on codeword lengths for small side information alphabet sizes are provided and by Koulgi, Tuncel and Rose [102] on theoretical properties of the achievable rate region.

Zero-error code constructions can be used to design near-lossless codes. This may be done by applying a zero-error construction on a modified joint probability distribution, where small entries are set to zero and the remaining entries are renormalized.

This is done explicitly in [193]. First, all the subsets of entries in the joint probability distribution that satisfy the given constraint on probability of confusion are listed. Then, for each of these subsets, a lossless MASC is designed with the modified joint probability distribution as input. Finally, the encoder with the minimum rate is selected. Although this approach results in an optimal code for the required maximum probability of confusion, such a construction is not practical. Heuristics may be used to speed up the search, at the cost of an increase in the rate R or the probability of confusion P_c.

Constructions based on graphs

Koulgi, Tuncel, Regunathan and Rose show an exponential-time optimal design algorithm for UI and RI codes based on confusability graphs [101]. They consider subgraphs induced by subsets of the symbol set $\mathcal{X}$. The optimal UI and RI rates are recursively related to those of the induced subgraphs.

In [103], the same team proposes a polynomial, yet suboptimal, design algorithm for good UI codes based on approximate graph colorings. It uses a coloring algorithm that gives close-to optimal results with respect to chromatic number, and then encodes the colors using a Huffman code.

Constructions based on syndromes

So far, all the code constructions were focused on the coding of individual symbols. With the construction of codes based on syndromes, the coding now processes a block of d symbols X.

A way for Claude to give Dominique information about X is to send the syndrome of a linear error correcting code $\alpha(X) = HX$, with X expressed in some vector space $\text{GF}(q)^d$ and H the parity check matrix of the code. Upon receiving $\xi = Hx$ for an outcome x of X, Dominique looks for the most probable $\bar{x}$ conditionally on $Y = y$ such that $H\bar{x} = \xi$.

Encoding and decoding in such constructions inherit from the standard techniques of coding theory. This technique is by itself so important that I will give two examples of such constructions in Sections 8.4 and 8.5.

8.3 Binary interactive error correction protocols

Binary interactive error correction protocols have traditionally been used for QKD protocols that produce binary key elements. They are also important for reconciliation in general, when combined with sliced error correction – see Section 9.1. I present here the existing protocols in a logical order.

X and Y are l-bit random variables $X, Y \in \text{GF}(2)^l$.

8.3.1 Bennett–Bessette–Brassard–Salvail–Smolin

The first binary interactive error correction (IEC) protocol used in the scope of QKD was designed by Bennett and coworkers [13] (BBBSS). It works on a long binary string and requires Claude and Dominique to exchange parities of subsets of their bits called subblocks. The presence of diverging parities help Claude and Dominique to focus on errors using a bisection and to correct them. BBBSS uses several iterations, between which the bit positions are permuted in a pseudo-random way.

In each iteration, Claude and Dominique divide the string into subblocks of (approximately) equal length called subblocks. For simplicity, assume that $l = nw$ for some integers n and w. The l-bit strings X and Y are cut into w-bit subblocks. For the first iteration, the subblocks are the sets of indices $B_j^{(1)} = \{(j-1)w + 1 \ldots jw\}$, for $1 \leq j \leq n$. For subsequent iterations, Claude and Dominique agree on a randomly-chosen permutation $\pi^{(i)}$. The subblocks contain the indices

$$B_j^{(i)} = \{\pi_t^{(i)} : (j-1)w + 1 \leq t \leq jw\}.$$

The parities of each of these subblocks are exchanged. For iteration i,

Claude and Dominique disclose the parities

$$\xi_{X,j}^{(i)} = \sum_{t \in B_j^{(i)}} X_t \text{ and } \xi_{Y,j}^{(i)} = \sum_{t \in B_j^{(i)}} Y_t,$$

respectively. When a subblock $B_j^{(i)}$ is such that $\xi_{X,j}^{(i)} \neq \xi_{Y,j}^{(i)}$, it means that there is an odd number of errors in the subblock, hence at least one. In such a case, a bisection begins: Claude and Dominique exchange the parity of half of the subblock. If the parity is wrong, they go on with the bisection in that half of the subblock; otherwise, at least one error is present in the other half of the subblock and the bisection focuses on that other half. The bisection ends when it has enclosed an erroneous bit. Knowing the position of this bit is enough for Dominique to correct it: s/he can simply flip it.

8.3.2 Cascade

Cascade [27] is an IEC based on BBBSS, but with an improved efficiency in terms of the number of disclosed bits. It uses four iterations. The first iteration of Cascade is identical to the first iteration of BBBSS, while the next three are different.

Unlike BBSSS, Cascade keeps track of all investigated subblocks and takes advantage of this information starting from the second iteration. More precisely, Cascade keeps two sets of subblocks: $\mathcal{B}_0$, the subblocks for which the parity is equal between Claude and Dominique, and $\mathcal{B}_1$, the subblocks with diverging parities. Each subblock B for which a parity was disclosed (including during the bisection) is listed in either $\mathcal{B}_0$ or $\mathcal{B}_1$.

When the bisection ends and an error is corrected (say bit b is flipped), Claude and Dominique go through the list of all the subblocks for which they already calculated the parity in the current and previous iterations. Any subblock containing bit b now sees its parity flipped due to the correction of b. Therefore, there may be subblocks for which the parity was equal between Claude and Dominique and which is now different. Let $\Delta_s \subseteq \mathcal{B}_s$, $s = 0, 1$, be the set of subblocks in $\mathcal{B}_s$ that contain the bit b. The sets of subblocks are updated in the following way:

$$\mathcal{B}_0 \leftarrow \mathcal{B}_0 \setminus \Delta_0 \cup \Delta_1,$$
$$\mathcal{B}_1 \leftarrow \mathcal{B}_1 \setminus \Delta_1 \cup \Delta_0.$$

Before the end of an iteration, Claude and Dominique correct all the known diverging parities. Among all the subblocks in $\mathcal{B}_1$, Claude and Dominique proceed with a bisection in the smallest of such subblocks. When they find an error to correct, they again update the parities of all the previous subblocks, update $\mathcal{B}_0$ and $\mathcal{B}_1$ and repeat the process until no known subblock has a diverging parity, i.e., until $\mathcal{B}_1 = \varnothing$.

The purpose of Cascade is to disclose as few parities as possible. In this perspective, the choice of subblock size w seems to play a critical role. If the subblock size w is large, say much larger than $1/e$, where $e = \Pr[X_i \neq Y_i]$ is the bit error rate, a large number of errors is, on average, contained in this subblock. Since a bisection is able to correct only one of them, an iteration will not be efficient. On the other hand, a small subblock size (i.e., much smaller than $1/e$) does not often contain an error. When Claude and Dominique reveal parities ξ_X and ξ_Y, they do not gain much information as $\xi_X = \xi_Y$ often occurs, or stated otherwise, $h(\Pr[\xi_X \neq \xi_Y]) \ll 1$. The eavesdropper, on the other hand, gains information for every parity revealed. The ideal situation would be that w is such that a subblock contains a small average number of errors while $h(\Pr[\xi_X \neq \xi_Y]) \approx 1$.

In Cascade, the subblock size can be chosen per iteration. The subblock sizes w_i, $i = 1\ldots 4$, must be chosen so as to globally reveal the least number of bits during the execution of Cascade while achieving the smallest possible probability of error between Claude's and Dominique's string at the end.

The subblock size in the original version of Cascade [27] is $w_1 \approx 0.73/e$ and $w_i = 2w_{i-1}$. An optimization of the subblock size was performed by Nguyen, who provides a list of optimal subblock sizes for a wide range of bit error rates e [135].

Other optimizations of Cascade concern the interleaving between two iterations. In Cascade, this permutation is chosen pseudo-randomly among all possible permutations. In the works of Chen [K. Chen, private communication (2001)] and of Nguyen [135], better interleaving methods are proposed to take the subblock structure into account and to try to avoid two erroneous bits being contained in the same subblock for the first two (or all four) iterations.

Let us briefly analyze the number of parities exchanged by Cascade. After the protocol, Claude and Dominique disclosed RX and RY for some matrix R of size $r \times l$. They thus communicated the parities calculated over identical subsets of bit positions. The matrix R and the number r of disclosed parities are not known beforehand but are the result of the various bisections and are a function of the number and positions of the diverging parities encountered. For the original Cascade [27], the number of parities r exchanged behaves roughly as $r/l \approx (1.1 + e)h(e)$. The optimization in [135] gives a number of parities close to

$$r/l \approx (1.0456 + 0.515e)h(e) + 0.0021. \tag{8.2}$$

The theoretical limit is $h(e)$.

Note that when $e = 25\%$, Cascade reveals as many parity bits as contained in the string, hence potentially disclosing all the information.

8.3.3 Furukawa–Yamazaki

Another IEC based on BBBSS is a protocol using perfect codes designed by Furukawa and Yamazaki [62] (FY). Like BBBSS, it also uses a certain number of iterations with bit interleaving in between.

Like BBBSS, FY also cuts the binary string into subblocks. Claude and Dominique exchange the parities of all their subblocks and thus determine which subblocks contain an odd number of errors. Instead of using an interactive bisection, the correction of the erroneous subblocks is one way from Claude to Dominique. For each subblock with a diverging parity, Claude sends Dominique the syndrome of a perfect code calculated over his/her subblock. Given this information, Dominique attempts to correct his/her subblock. (Notice the similarity with the syndromes as in Section 8.2.4.)

Unfortunately, this protocol is less efficient than Cascade in terms of the number of bits disclosed. However, the underlying idea is interesting and a more efficient protocol based on identical principles is studied next.

8.3.4 Winnow

The Winnow protocol [31] is an IEC very similar to FY. Note that Winnow also includes a privacy amplification-like step that discards bits during the error correction. We, however, do not take this aspect into account here. Like the other IECs so far, it uses a certain number of iterations with bit interleaving in between.

Like BBBSS, FY and Cascade, Winnow also cuts the binary string into subblocks. Claude and Dominique exchange the parities of all their subblocks and thus determine which subblocks contain an odd number of errors. For the subblocks with diverging parity, Claude sends Dominique the syndrome of a Hamming code calculated over his/her subblock.

Unlike BBBSS and Cascade, which use a bisection, the correction of a subblock using the Hamming code does not necessarily reduce the number of errors in that subblock. The Hamming code proposed in Winnow allows Claude and Dominique to correct one error. If more than one error is present in the subblock, Dominique's attempt may actually increase the number of errors in that subblock. The subblock size should be chosen in such a way that it globally reduces the number of errors.

An optimization of the subblock sizes for Winnow was also performed by Nguyen [135]. Unlike Cascade, the iterations of Winnow are independent of each other and so an exhaustive search could be performed at a low complexity using dynamic programming [44].

The cost of Winnow as a function of the bit error rate does not follow a nice curve, as Cascade does. Consequently, we rely on the analysis in [135] to draw some conclusions. Cascade performs better than Winnow for bit error rates up to about 10%. Between 10% and 18%, Winnow is more efficient. Winnow does not work properly above 18%, so Cascade has to be used again, until 25%, an error rate at which Cascade has to reveal the entire string to be able to correct all errors.

8.3.5 Interactivity levels of Cascade and Winnow

The two most efficient IECs being Cascade and Winnow, let us compare their requirements in terms of interactivity.

An important difference between Cascade and Winnow is the level of interactivity needed. During the execution of Cascade – at least for the second iteration and further – Claude has to wait for a parity bit from Dominique before she knows which parity bit she has to transmit, and vice versa. They may be doing a bisection and they have to wait to see whether they have equal or different parities to decide to focus on the left or on the right half of the current subblock. Each party can thus only transmit one bit at a time.

This one-bit interactivity can be a problem in practice. First, depending on the network services used, a one-bit message has to be included in a much larger message of hundreds or thousands of bytes. From a data compression point of view, it is difficult to make this less efficient. Then, the number of messages depends on the size of the string to reconcile. The transmission of a message may suffer from a latency time, which has to be multiplied by the number of parity bits to exchange. The latency time is thus proportional to l.

A way to avoid this huge loss in transmission efficiency is to divide the l-bit string to reconcile into chunks of l' bits, where l' is a fixed parameter, e.g., $l' = 10\,000$. Each chunk is reconciled by an independent instance of Cascade, requiring thus about $\nu = l/l'$ parallel instances of the protocol. The execution of the instances are synchronized in such a way that the $\nu \times 1$

bits can be transmitted in a single message. The number of messages, and thus the latency time, is no longer proportional to l.

The case of Winnow is quite different. The parities of all subblocks can be transmitted in a single message, from Claude to Dominique and then from Dominique to Claude. Then, the Hamming syndromes of all the erroneous subblocks can be sent in the same message from Claude to Dominique. Winnow thus requires only three messages per iteration, independently of the number of bits to reconcile.

8.4 Turbo codes

When used for channel coding or for source coding with side information, turbo codes have been found to achieve rates very close to the Shannon limit. The good performances of turbo codes is mainly due to the use of iterated soft decoding, that is, where the decoding does not only yield a decoded bit but also associates a confidence to it.

Turbo codes have unprecedented performances and the original paper by Berrou, Glavieux and Thitimajshima [17] started a revolution in the information theory community. Turbo codes have been carefully analyzed and have received many improvements. Also, other kinds of powerful codes using iterated soft decoding techniques were found, such as low-density parity-check (LDPC) codes, which I will introduce in Section 8.5.

In this section, I give a short introduction to turbo codes with a clear focus on source coding with side information instead of channel coding [1]. First, I describe the convolutional codes, as they constitute an essential ingredient of turbo codes. Then, I explain how these codes can be soft-decoded. Finally, these ingredients are assembled to make the turbo codes.

8.4.1 Convolutional codes

Unlike traditional error correcting codes, which work on blocks of symbols of particular sizes, the encoder of a convolutional code takes as its input a stream of bits and outputs a stream of bits. For practical reasons, we assume that it encodes a l-bit string, but this value can be freely specified by the user of the code without any essential change in the code structure.

The convolutional encoder contains an m-bit state $s = (s^{(1)}, \ldots, s^{(m)})$, which evolves as a function of the input bits. The output stream is a linear function of both the input bits and of the state. The encoder is time independent and can be seen as a convolutional filter in $GF(2)$, hence its name. (Notice that the convolutional encoder shares some similarity with linear feedback shift registers; see Section 2.1.2.)

We restrict ourselves more specifically to *binary systematic recursive convolutional codes*. In this scope, *systematic* means that the input bits appear unaltered in the output stream. The encoder is *recursive* because the content of the state is fed back into itself and into the output stream.

As depicted in Fig. 8.3, the output of the convolutional code consists of two streams of bits: one that contains the input bits $x_{1...l}$ unaltered, called the *systematic bits*, and one that contains *parity bits* $\xi_{1...l}$.

The convolutional code is defined by the formal ratio of two polynomials in D:

$$G(D) = f(D)/g(D),$$
$$f(D) = f_0 + f_1 D + \cdots + f_m D^m,$$
$$g(D) = g_0 + g_1 D + \cdots + g_m D^m,$$

where it is assumed that $g_0 = 1$. The symbol D must be thought of as a delay in the encoder's memory. The polynomial f indicates how the parity bits are generated from the state, whereas the polynomial g specifies how the state evolves and is fed back. As made more precise below, the coefficients f_j and g_j indicate whether there is a connection from jth state bit to the parity and to the feedback, respectively. Another conventional way to define a convolutional code is to evaluate the polynomials at $D = 2$ and to write the ratio of the two resulting integers. For instance, we talk about the convolutional code 7/5 for $G(D) = (1 + D + D^2)/(1 + D^2)$.

Let us now describe how the convolutional encoder works for a given $G(D)$. For each incoming input bit x_t at time t, $1 \le t \le l$, the parity bits and state bits evolve in the following way:

$$\xi_t = f_0 x_t + \sum_{j=1...m} f_j s_{t-1}^{(j)} + f_0 g_m s_{t-1}^{(m)},$$
$$s_t^{(1)} = x_t + \sum_{j=1...m} g_j s_{t-1}^{(j)},$$
$$s_t^{(i)} = s_{t-1}^{(i-1)} \text{ for } 2 \le i \le m.$$

Before the first iteration, the state is set to zero, that is, $s_0^{(j)} = 0$ for $1 \le j \le m$.

Turbo codes, and therefore convolutional codes, are originally defined for channel coding. For source coding with side information, two minor adaptations are required, as we now describe.

First, channel coding requires that both the systematic and the parity bits

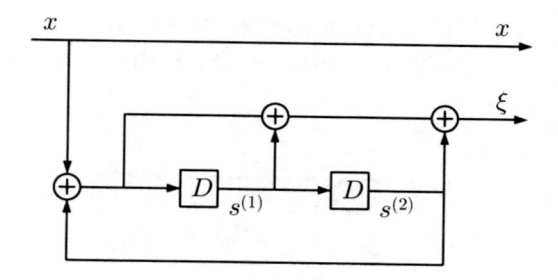

Fig. 8.3. Example of a convolutional encoder with an $m = 2$-bit memory. This encoder is specified by $G(D) = (1+D+D^2)/(1+D^2)$ or 7/5. There are connections from the state bits $s^{(1)}$ and $s^{(2)}$ to the parity bits as the numerator contains D and D^2, respectively. Notice that there is no connection from $s^{(1)}$ to the feedback as the coefficient of D in the denominator is 0.

are sent over the channel, as both are needed by the recipient to recover the original message. For source coding with side information, however, only the parity bits are sent as side information over the (noiseless) public classical authenticated channel. A noisy version of the systematic bits is known to Bob through Y, as if X was sent over a noisy channel. Of course, not all parity bits are sent as side information. As the goal of source coding with side information is to send the least number of bits so that Bob can recover X, only a well-chosen subset of the parity bits are kept. The process of removing parity bits is called puncturing, and we will come back to it later, when we describe the turbo codes as such.

Second, when turbo codes are used for channel coding, the encoding of the input string outputs an additional m parity bits $\xi_{l+1...l+m}$ after the encoding of the l input bits. These last parity bits are produced with input bits equal to 0 and no state feedback, so as to force the final state to be $s = (0, 0, \ldots, 0)$. This fixes both the initial and the final states as boundary conditions, which are required by the decoder. In the scope of source coding with side information, we instead assume that Alice reveals the final state as side information, i.e., she gives Bob $\sigma_l = s_l$. This also fixes both the initial and the final states in an equivalent, but easier, way.

A popular way to represent convolutional codes graphically is to use *trellis*. Informally, a trellis is a bipartite graph with a set of nodes that represents the 2^m possible states at time $t-1$, or input states, and a set of nodes that represents the states at time t, or output states. The trellis comprises edges that connect state s_{t-1} to state s_t if and only if there is an input symbol x_t that causes a state transition from s_{t-1} to s_t. The edges are labeled with the associated input bit x that causes this particular transition and with the associated parity bit ξ which is output in these circumstances. Note that the encoding is independent of t and so is the trellis. An example of a trellis is given in Fig. 8.4.

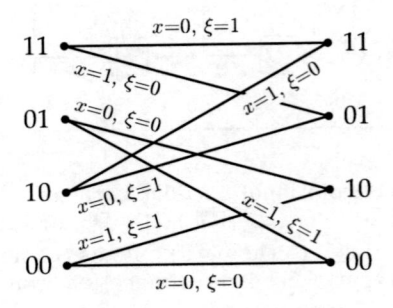

Fig. 8.4. A trellis for the convolutional code 7/5. The input states are located on the left and are denoted as $s^{(1)}s^{(2)}$. The output states are located on the right.

Note that the trellis can be concatenated so as to represent the state transitions from s_0 to s_l. For a given input string, the state transitions make a path from the node $s_0 = 0$ to the node $s_l = \sigma_l$, as illustrated in Fig. 8.5.

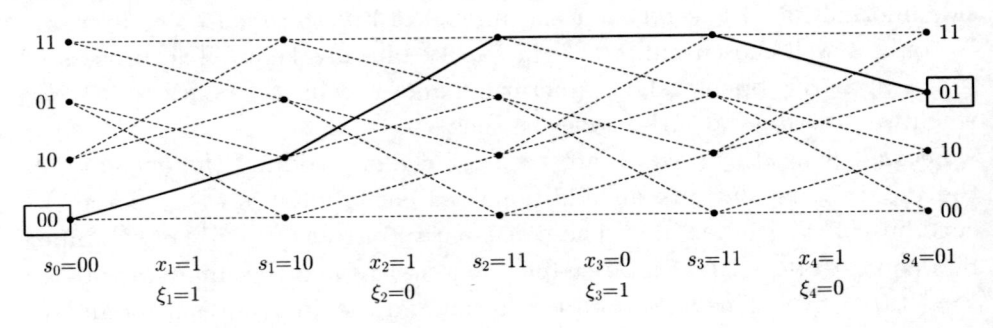

Fig. 8.5. An example of the state transition path for the convolutional code 7/5 for the $l = 4$-bit input block $x_{1...4} = 1101$. The initial state is $s_0 = 00$ and the final state is $\sigma_4 = s_4 = 01$.

8.4.2 Maximum a-posteriori decoding of convolutional codes

Now that we have described the encoder, let us describe how Bob can recover X from Y and ξ. Note that we are not yet talking about turbo codes themselves, only about the convolutional codes. In fact, as a constituting part of the turbo codes, we wish to recover from Y and ξ not only the value of a bit but also an estimate of the probability of it being a zero or a one. This is the soft decoding required by the turbo codes.

The maximum a-posteriori (MAP) algorithm described here was invented by Bahl and coworkers [5]. The goal is to find, for each symbol x_t, which one is the most probable *a posteriori*, that is, when all the symbols $y_{1...l}$ have been received.

As a convention in soft decoding, we estimate the *log-likelihood ratio* (LLR) of the a-posteriori probabilities, i.e.,

$$L(X_t|y_{1...l}) = \ln \frac{\Pr[X_t = 0|Y_{1...l} = y_{1...l}]}{\Pr[X_t = 1|Y_{1...l} = y_{1...l}]} = \ln \frac{\Pr[X_t = 0, Y_{1...l} = y_{1...l}]}{\Pr[X_t = 1, Y_{1...l} = y_{1...l}]}.$$

The value of $L = L(X_t|y_{1...l})$ is positive if $X_t = 0$ is more likely than $X_t = 1$, and vice versa if the value is negative. If $L = 0$ both values are equally likely; when $X_t = 0$ (or $X_t = 1$) is certain, we have $L = +\infty$ (or $L = -\infty$).

Instead of looking only at the symbols x_t and $y_{1...l}$, we evaluate the LLR by calculating the probabilities of the different state transitions. In particular, we look at the state transition at time t, for all paths going from $s_0 = 0$ to s_{t-1} and all paths starting in s_t and ending in $s_l = \sigma_l$, for all possible values of s_{t-1} and s_t. The LLR can, equivalently, be written as

$$L(X_t|y_{1...l}) = \ln \frac{\sum_{s_{t-1},s_t,\xi_t} P(s_{t-1}, s_t, 0, \xi_t)}{\sum_{s_{t-1},s_t,\xi_t} P(s_{t-1}, s_t, 1, \xi_t)}, \qquad (8.3)$$

where $P(s_{t-1}, s_t, x_t, \xi_t)$ is a shorthand notation for

$$P(s_{t-1}, s_t, x_t, \xi_t) = \Pr[S_{t-1} = s_{t-1}, S_t = s_t, Y_{1...l} = y_{1...l}, X_t = x_t, \Xi_t = \xi_t].$$

Let us first expand all the terms and then describe the different factors one by one. The probability $P(s_{t-1}, s_t, x_t, \xi_t)$ can be rewritten as

$$P(s_{t-1}, s_t, x_t, \xi_t) = \alpha_{t-1}(s_{t-1})\gamma(s_{t-1}, s_t, x_t, \xi_t)\beta_t(s_t),$$
$$\alpha_t(s_t) = \Pr[S_t = s_t, Y_{1...t} = y_{1...t}],$$
$$\gamma(s_{t-1}, s_t, x_t, \xi_t) = \delta(s_{t-1}, s_t, x_t, \xi_t)P_{Y|X}(y_t|x_t)P_X(x_t)P_\Xi(\xi_t),$$
$$\beta_t(s_t) = \Pr[S_t = s_t, Y_{t+1...l} = y_{t+1...l}],$$
$$\delta(s_{t-1}, s_t, x_t, \xi_t) = \Pr[S_t = s_t, \Xi_t = \xi_t|S_{t-1} = s_{t-1}, X_t = x_t].$$

The function $P(s_{t-1}, s_t, x_t, \xi_t)$ is first divided into three factors. First, the function $\alpha_t(s_t)$ looks at the past and tells us how likely it is that the encoder arrives at state s_t given that we observed $y_{1...t}$. Then, the function $\beta_t(s_t)$ looks at the future and, knowing that the decoder arrives in state σ_l at time l, gives us the probability that we started in state s_t at time t. Finally, the function $\gamma(s_{t-1}, s_t, x_t, \xi_t)$ tells us how likely it is that a transition occurs between s_{t-1} and s_t at time t.

As a part of the γ function, $\delta(s_{t-1}, s_t, x_t, \xi_t)$ is a function that returns 1 if the transition from s_{t-1} to s_t with emitted parity ξ_t exists for x_t as input, and 0 otherwise. So, even though we sum over all past and future states in Eq. (8.3), this function δ ensures that only the valid transitions are selected in the sum.

The probability $P_{Y|X}(y_t|x_t)P_X(x_t) = P_{XY}(x_t, y_t)$ is crucial for the transition at time t, as it explicitly takes into account the noisy value of X_t that Bob received in Y_t. Here, $P_X(x_t)$ is known as the *a-priori information* on X_t. By default, $P_X(x_t) = 1/2$, but as we will see in more detail in Section 8.4.3 iterative decoding of turbo codes makes this value evolve.

Depending on whether the parity bit was punctured at time t, the probability $P_\Xi(\xi_t)$ is either $P_\Xi(\xi_t) = 1/2$ if no parity bit was received or $P_\Xi(\xi_t) = 0$ or 1 depending on the received parity bit value. (Remember that in the case of source coding with side information, the parity bits are transmitted losslessly.) So, when a parity bit is known, the sum in Eq. (8.3) takes only into account the state transitions that match the received parity bit.

To sum up, the γ function takes into account the possible state transitions and depends on the a-priori information and on the actual noisy values and parity bits received by Bob. To proceed with the MAP algorithm, γ is evaluated for all times $1 \le t \le l$ and for all possible state transitions. These values give local information on the likelihood of a given state transition at a given time. To compute the LLRs, however, the values of γ must be combined so as to take into account the global picture of all possible state transition paths. This is where the α and β functions are also needed. Actually, α and β can be both efficiently evaluated in a recursive way, combining the values of γ at different times.

The function $\alpha_t(s_t)$ verifies the property that

$$\alpha_t(s_t) = \sum_{x_t, \xi_t, s_{t-1}} \gamma(s_{t-1}, s_t, x_t, \xi_t) \alpha_{t-1}(s_{t-1}).$$

The values of $\alpha_t(s_t)$ for all t and all s_t can thus be computed by starting from $t = 1$, with the convention that $\alpha_0(0) = 1$ and $\alpha_0(s) = 0$ for $s \ne 0$.

Similarly, the function $\beta_t(s_t)$ can be computed recursively starting from the last state. The recursion reads

$$\beta_{t-1}(s_{t-1}) = \sum_{x_t, \xi_t, s_t} \gamma(s_{t-1}, s_t, x_t, \xi_t) \beta_t(s_t).$$

The values of $\beta_t(s_t)$ can be computed by starting from $t = l$, with the convention that $\beta_l(\sigma_l) = 1$ and $\beta_l(s_l) = 0$ for $s_l \neq \sigma_l$.

We now describe the structure of turbo codes as such.

8.4.3 Encoding and decoding of turbo codes

As depicted in Fig. 8.6, turbo codes consist of two (usually identical) convolutional codes operating in parallel. Before entering the second convolutional encoder, the input bits are shuffled using an interleaver. The interleaving often takes the form of a pseudo-random permutation and spreads the bits so that the second encoder produces a different family of parity bits.

As usual for source coding with side information, the systematic bits are discarded. The $2l$ parity bits of both encoders are punctured so as to keep only a fraction of them. The number of parity bits kept depends on the information already shared by Alice and Bob and should be close to $lH(X|Y)$. Note that different puncturing strategies exist, such as discarding a pseudo-random subset of the parity bits.

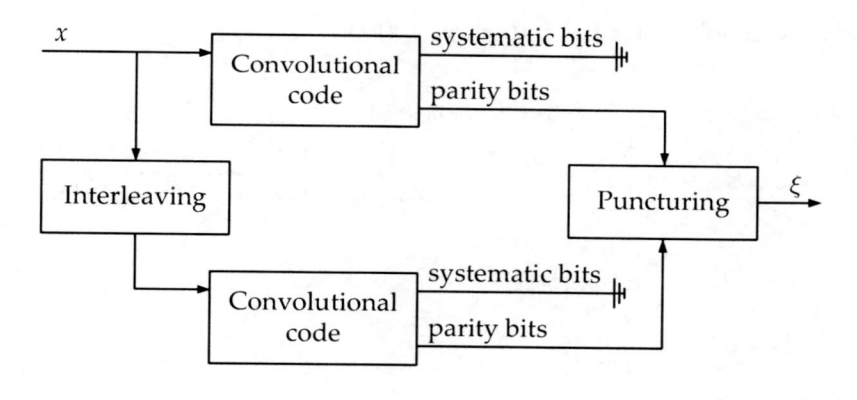

Fig. 8.6. Structure of a turbo encoder.

The decoding of turbo codes relies on the soft decoding of both convolutional encoders. The good performance of turbo codes comes from the fact that each convolutional decoder takes advantage of the soft decoding of the other. Let us describe this process in more detail.

As a start, the first convolutional code is decoded using the MAP algorithm with the parity bits ξ_{CC1} produced by the first encoder. This process yields the LLRs $L_1(X_t|y_{1...l})$, $1 \leq t \leq l$, as described in Section 8.4.2. For this first decoding, the a-priori probabilities $P_X(x_t)$ are initialized to $P_X(x_t) = 1/2$ since both bits are equally likely.

Before processing the parity bits ξ_{CC2} in the second decoder, however, the a-priori probabilities $P_X(x_t)$ are initialized as a function of $L_1(X_t|y_{1...l})$. In this way, the knowledge of decoder 1 is passed on to decoder 2. Then, the LLRs produced by decoder 2, $L_2(X_t|y_{1...l})$, are passed on to decoder 1 as a-priori probabilities. The parity bits ξ_{CC1} are thus processed a second time, yielding finer results, which are passed on to decoder 2, and so on. This alternative process is repeated until convergence is reached or for fixed number of iterations.

Let us take a closer look at the LLRs. In fact, the numerator and denominator of Eq. (8.3) can be factored as

$$\sum_{s_{t-1},s_t,\xi_t} P(s_{t-1}, s_t, x_t, \xi_t) = \left(\sum_{s_{t-1},s_t,\xi_t} \alpha_{t-1}(s_{t-1})\delta(s_{t-1}, s_t, x_t, \xi_t) P_\Xi(\xi_t)\beta_t(s_t) \right) \times P_{Y|X}(y_t|x_t) \times P_X(x_t).$$

Thus, we can split the LLR into three terms,

$$L(X_t|y_{1...l}) = L_{\text{ext}} + L_{\text{ch}} + L_{\text{a-priori}}, \text{ with}$$

$$L_{\text{ext}} = \ln \frac{\sum_{s_{t-1},s_t,\xi_t} \alpha_{t-1}(s_{t-1})\delta(s_{t-1}, s_t, 0, \xi_t) P_\Xi(\xi_t)\beta_t(s_t)}{\sum_{s_{t-1},s_t,\xi_t} \alpha_{t-1}(s_{t-1})\delta(s_{t-1}, s_t, 1, \xi_t) P_\Xi(\xi_t)\beta_t(s_t)},$$

$$L_{\text{ch}} = \ln \frac{P_{Y|X}(y_t|0)}{P_{Y|X}(y_t|1)},$$

$$L_{\text{a-priori}} = \ln \frac{P_X(0)}{P_X(1)}.$$

The LLR given as output by the MAP algorithm for one encoder consists of two terms that can actually be determined before running the algorithm: L_{ch}, which depends only on the correlations between X and Y (i.e., on the channel in the case of channel coding), and $L_{\text{a-priori}}$, which is given at the input of the MAP algorithm. These two terms are not used in the exchange between the two decoders.

In contrast, the last term, L_{ext}, produced by one encoder contains a value that depends only on the information unknown to the other decoder, called the *extrinsic information*, and that can be passed on as an a-priori probability to the other decoder. Given the value L_{ext} returned by the MAP algorithm of decoder 1 (or decoder 2), the a-priori probabilities for decoder

2 (or decoder 1) are initialized as

$$P_X(x) = \frac{e^{(-1)^x L_{\text{ext}}/2}}{e^{L_{\text{ext}}/2} + e^{-L_{\text{ext}}/2}}.$$

8.5 Low-density parity-check codes

The low-density parity-check (LDPC) codes were first discovered in 1962 by Gallager [63]. They were forgotten for some time until recently, when they have been rediscovered and raised new interests.

An LDPC code is an error-correcting code determined by a particular form of a $r \times l$ parity-check matrix $H \in \text{GF}(2)^{rl}$. More specifically, a family of LDPC codes is characterized by the proportion of non-zero entries in each row and in each column, as summarized by the two formal polynomials

$$L(x) = \sum_k L_k x^k \text{ and } R(x) = \sum_k R_k x^k.$$

The polynomial $L(x)$ (or $R(x)$) indicates the proportion of non-zero entries in the columns (or rows) of H, i.e.,

$$L_k = l^{-1}|\{j : \text{column } H_{\cdot j} \text{ contains } k \text{ ones}\}|,$$
$$R_k = r^{-1}|\{i : \text{row } H_{i \cdot} \text{ contains } k \text{ ones}\}|.$$

For an LDPC code with polynomials $L(x)$ and $R(x)$, the parity check matrix thus contain a total of $E(l) = l \sum_k k L_k = r \sum_k k R_k$ non-zero entries. This number $E(l)$ grows only proportionally to the block size l, whereas a randomly-chosen matrix H would contain a quadratic number of non-zero entries (i.e., assuming that r is proportional to l). This is the reason for the qualifier *low-density* of LDPC codes.

With the matrix size fixed, all the LDPC codes characterized by the same polynomials $L(x)$ and $R(x)$ have pretty much the same properties. Hence, it is sufficient for our purposes to think of an LDPC code as being randomly chosen among an ensemble; with overwhelming probability, such a code will be as good as any other code in the ensemble.

Since we wish to use LDPC codes for reconciliation, we focus on the source coding with side information problem – see also the works of Liveris, Xiong and Georghiades and of Muramatsu and coworkers [107, 129, 130]. For encoding, Alice sends Bob the syndrome of her key elements, i.e., she sends $\Xi = HX$. Note that this operation requires only $E(l)$ binary operations, which is proportional to the block size l.

For decoding, Bob uses an iterative process based on LLRs in a way similar to turbo codes. Before detailing the decoding procedure, it is convenient to describe the LDPC codes as *Tanner graphs*. For a given LDPC code, let its associated Tanner graph $G(H)$ have the set of $l + r$ vertices $v_{1...l}$ and $c_{1...r}$, called variables nodes and check nodes, respectively. The edges of the Tanner graph are determined by the matrix H: an edge connects v_j to c_i iff $H_{ij} = 1$. Hence, the Tanner graph is a bipartite graph with $E(l)$ edges.

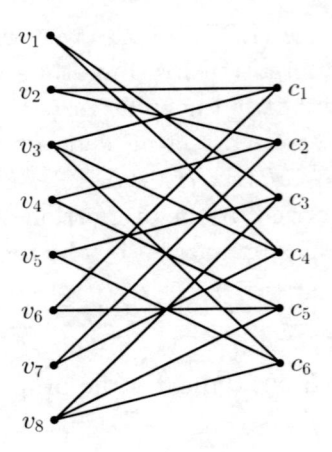

Fig. 8.7. The Tanner graph of some arbitrary 6×8 LDPC code with polynomials $L(x) = \frac{6}{8}x^2 + \frac{2}{8}x^3$ and $R(x) = \frac{6}{6}x^3$.

From a graph perspective, the polynomials $L(x)$ and $R(x)$ give the degrees of the variable nodes and of the check nodes, respectively. In particular, lL_k variable nodes are of degree k, while rR_k check nodes are of degree k.

The decoding process can be described as an exchange of messages between variable and check nodes. In particular, let us now describe the *belief propagation* decoding algorithm. Bob applies a procedure along the following lines. First, the variable nodes send the noisy values y to the adjacent check nodes. The check nodes should receive values from the variable nodes that sum to ξ. Based on this assertion, the check nodes then send back to the variable nodes what they "think" is the correct value of x. Since y is noisy, this may be equal to or different from the values first sent. At this point, the variable nodes receive the opinion of the different check nodes, which give them a more accurate value for x (i.e., the LLR of X). This new value is propagated again to the check nodes, and so on.

To help describe the exact form of the procedure, let us first play with a toy example. Assume that Bob already knows the correct value x sent by Alice. As part of the message passing algorithm, each variable node v_j first sends the value of the corresponding bit x_j as message to the adjacent check node c_i (i.e., for $H_{ij} = 1$ so that an edge between v_j and c_i exists). From the perspective of node c_i, all the received messages must sum up to ξ_i since $\xi_i = \sum_j H_{ij} x_j$. Then, the check node c_i sends to variable node v_j the value $\xi_i + \sum_{j' \neq j} H_{ij'} x_{j'}$. From the perspective of variable node v_j, it should receive the value x_j from all the adjacent check nodes since $\xi_i + \sum_{j' \neq j} H_{ij'} x_{j'} = x_j$.

Of course, Bob only has access to the parity bits ξ and to his own key elements y, which are noisy versions of the key elements x he wishes to recover. The belief propagation algorithm translates the toy example above in terms of LLRs. For the first step, the variable node v_j knows only y_j. Bob evaluates the associated LLR

$$L_{v_j \to c} = L(X_j | y_j) = \ln \frac{\Pr[X_j = 0, Y_j = y_j]}{\Pr[X_j = 1, Y_j = y_j]},$$

which is propagated to the check nodes. Then, the LLR of the expression $\xi_i + \sum_{j' \neq j} H_{ij'} x_{j'}$ must be propagated back to the variable node v_j. To see how it works, let us evaluate the LLR of the sum of two bits $X_1 + X_2$:

$$\begin{aligned} L(X_1 + X_2 | y_1, y_2) &= \ln \frac{P_{X_1+X_2,Y_1,Y_2}(0, y_1, y_2)}{P_{X_1+X_2,Y_1,Y_2}(1, y_1, y_2)} \\ &= \ln \frac{P_{X_1 Y_1}(0, y_1) P_{X_2 Y_2}(0, y_2) + P_{X_1 Y_1}(1, y_1) P_{X_2 Y_2}(1, y_2)}{P_{X_1 Y_1}(0, y_1) P_{X_2 Y_2}(1, y_2) + P_{X_1 Y_1}(1, y_1) P_{X_2 Y_2}(0, y_2)} \\ &= \ln \frac{\frac{P_{X_1 Y_1}(0, y_1)}{P_{X_1 Y_1}(1, y_1)} \times \frac{P_{X_2 Y_2}(0, y_2)}{P_{X_2 Y_2}(1, y_2)} + 1}{\frac{P_{X_1 Y_1}(0, y_1)}{P_{X_1 Y_1}(1, y_1)} + \frac{P_{X_2 Y_2}(0, y_2)}{P_{X_2 Y_2}(1, y_2)}} \\ &= \ln \frac{e^{L_1} e^{L_2} + 1}{e^{L_1} + e^{L_2}} = \ln \frac{\frac{e^{L_1}+1}{e^{L_1}-1} \frac{e^{L_2}+1}{e^{L_2}-1} + 1}{\frac{e^{L_1}+1}{e^{L_1}-1} \frac{e^{L_2}+1}{e^{L_2}-1} - 1} \\ &= \phi^{-1}(\phi(L_1)\phi(L_2)), \end{aligned}$$

with $L_i = L(X_i | y_i)$ and

$$\phi(L) = \frac{e^L + 1}{e^L - 1}, \quad \phi^{-1}(\lambda) = \ln \frac{\lambda + 1}{\lambda - 1}.$$

The generalization to the sum of several bits is straightforward. Hence, the LLR of the expression $\xi_i + \sum_{j' \neq j} H_{ij'} x_{j'}$ reads

$$L_{c_i \to v_j} = (-1)^{\xi_i} \phi^{-1} \left(\prod_{j' \neq j : H_{ij'}=1} \phi(L_{v_{j'} \to c}) \right).$$

Note that the $(-1)^{\xi_i}$ factor results from the fact that ξ is known with certainty and that $\phi^{-1}(\phi(\pm \infty) \phi(L)) = \pm L$.

The variable nodes receive LLRs from all the adjacent check nodes. How can Bob refine his estimation of the LLR of X using this information? For this, the belief propagation algorithm processes the incoming LLRs as if coming from independent observations. Assume that some variable X' is observed through the variables $Y_{1\ldots n}$ such that

$$\Pr[Y_{1\ldots n} = y_{1\ldots n} | X' = x'] = \prod_i \Pr[Y_i = y_i | X' = x'].$$

Then,

$$L(X'|y_{1\ldots n}) = \ln \frac{\Pr[Y_{1\ldots n} = y_{1\ldots n} | X' = 0]}{\Pr[Y_{1\ldots n} = y_{1\ldots n} | X' = 1]}$$
$$= \ln \frac{\prod_i \Pr[Y_i = y_i | X' = 0]}{\prod_i \Pr[Y_i = y_i | X' = 1]} = \sum_i L(X'|y_i).$$

So, under the assumption of independent observations, the LLRs can be summed. In the case of x_j, the variable node v_j takes into account both the LLR $L(X_j|y_j)$ and the LLR given by the adjacent check nodes $L_{c_i \to v_j}$. Hence, the LLR propagation rule at variable node reads

$$L_{v_j \to c} = L(X_j|y_j) + \sum_i L_{c_i \to v_j}.$$

This process is repeated until convergence or for a fixed number of iterations.

8.6 Conclusion

In this chapter, I explained how the information shared by Alice and Bob (or Claude and Dominique) can be reconciled to produce a common bit string. The difficult part of the problem is to make this process efficient in terms of disclosed information, as the secret key rate is highly dependent on this efficiency.

After an introduction to the techniques of source coding with side information, we showed how to use them for reconciliation. We also introduced several interactive and one-way protocols for reconciling binary strings.

These binary reconciliation protocols can be used for the bits produced by the BB84 protocol. Some QKD protocols produce continuous key elements, for which the next chapter presents suitable reconciliation techniques, thereby generalizing the techniques presented in this chapter.

9
Non-binary reconciliation

Some QKD protocols, as I will detail in Chapter 11, produce Gaussian key elements. The reconciliation methods of the previous chapter are, as such, not adapted to this case. In this chapter, we build upon the previous techniques to treat the case of continuous-variable key elements or, more generally, the case of non-binary key elements.

In the first two sections, I describe two techniques to process non-binary key elements, namely sliced error correction and multistage soft decoding. Then, I conclude the chapter by giving more specific details on the reconciliation of Gaussian key elements.

9.1 Sliced error correction

Sliced error correction (SEC) is a generic reconciliation protocol that corrects strings of non-binary elements using binary reconciliation protocols as primitives [173]. The purpose of sliced error correction is to start from a list of correlated values and to give, with high probability, equal binary strings to Claude and Dominique. The underlying idea is to convert Claude's and Dominique's values into strings of bits, to apply a binary correction protocol (BCP) on each of them and to take advantage of all available information to minimize the number of exchanged reconciliation messages. It enables Claude and Dominique to reconcile a wide variety of correlated variables X and Y while relying on BCPs that are optimized to correct errors modeled by a binary symmetric channel (BSC).

An important application of sliced error correction is to correct correlated Gaussian random variables, namely $X \sim N(0, \Sigma)$ and $Y = X + \epsilon$ with $\epsilon \sim N(0, \sigma)$. This important particular case is needed for QKD protocols that use a Gaussian modulation of Gaussian states, as described in Chapter 11.

9.1.1 Definitions

The random variables X and Y are defined over the sets $\mathcal{X}$ and $\mathcal{Y}$. To remain general, Claude and Dominique are free to group d variables at a time and to process d-dimensional vectors. Here, X and Y denote d-dimensional variables, taking values in $\mathcal{X}^d$, with $\mathcal{X}^d = \mathbf{R}^d$ in the particular case of Gaussian variables. For simplicity, we usually do not write the dimension d. When explicitly needed by the discussion, however, the dimension of the variables is noted with a $\cdot^{(d)}$ superscript.

To define the protocol, we must first define the *slice* functions. A slice $S(x)$ is a function from $\mathcal{X}$ to $\{0,1\}$. The slices $S_{1...m}(x) = (S_1(x), \ldots, S_m(x))$ are chosen so as to map Claude's key elements to a discrete alphabet of size at most 2^m, and are used to convert Claude's key elements into binary digits, that is,

$$\Psi(X) = S_{1...m}(X).$$

We proceed with the definition of *slice estimators*. Each of the slice estimators $E_1(y), \ldots, E_i(y, s'_{1...i-1}), \ldots, E_m(y, s'_{1...m-1})$ defines a mapping from $\mathcal{Y} \times \{0,1\}^{i-1}$ to $\{0,1\}$. These are used by Dominique to guess $S_i(X)$ as best as possible given his/her knowledge of Y and of the previously corrected slice bits $s'_{1...i-1}$ of lower indexes.

The construction of the slices $S_i(X)$ and their estimators depends on the nature and distribution of X and Y. These aspects are covered in a following section, where we apply the SEC to Gaussian key elements.

Let us now describe the generic protocol, which assumes that the legitimate parties defined and agreed on the functions S_i and E_i. Claude and Dominique also choose a block length l, independently of d, so as to process the l key elements $x_{1...l}$ and $y_{1...l}$. We assume that the l values x_j (or y_j) are independent outcomes of X (or Y) for different subscripts j.

The protocol goes as follows. For $i = 1$ to m, successively, Claude and Dominique perform the following steps:

- Claude prepares the string of bits $(S_i(x_1), \ldots, S_i(x_l))$.
- Dominique prepares the string of bits

$$(E_i(y_1, S_{1...i-1}(x_1)), \ldots, E_i(y_l, S_{1...i-1}(x_l))),$$

where $S_{1...i-1}(x_1)$ is known to Dominique, with high probability, from the previous $i-1$ steps.

- Claude and Dominique make use of a chosen BCP so that Dominique acquires the knowledge of Claude's bits $(S_i(x_1), \ldots, S_i(x_l))$.

9.1.2 Properties of sliced error correction

Disclosed information

The goal of reconciliation is for Claude and Dominique to obtain common bits (i.e., $l \times m$ bits $\Psi(x_j) = S_{1...m}(x_j)$, $j = 1...l$) by disclosing as little information as possible about them. However, one does not expect a protocol running with strings of finite length and using finite computing resources to achieve a net key rate equal to $I(X;Y)$ exactly. Yet it is easy to show that SEC is indeed asymptotically efficient, that is, it reaches the Slepian–Wolf bound in terms of leaked information when the number of dimensions d goes to infinity.

In the context of SEC, Corollary 1 of Section 8.1.2 means that, with d sufficiently large, there exist slice functions such that disclosing the first

$$r = \lfloor dH(\Psi(X^{(1)})|Y^{(1)}) + 1 \rfloor$$

slices $S_{1...r}(X^{(d)})$ is enough for Dominique to recover the $m - r$ remaining ones and to reconstruct $S_{1...m}(X^{(d)})$ with arbitrarily low probability of error. The variable Y plays the role of side information as usual, and the r bits produced by the r first slices form an encoding of X.

For continuous variables Y, it is necessary here to quantize Y, as Slepian and Wolf's theorem assumes discrete variables. In fact, Y can be approximated as accurately as necessary by a discrete variable $T(Y)$, with

$$H(\Psi(X)|T(Y)) \to H(\Psi(X)|Y).$$

For the practical case of a fixed dimension d, let us now analyze the amount of information leaked on the public channel during the processing of sliced error correction. Clearly, this depends on the primitive BCP chosen. This aspect will be detailed in a following section.

If not using SEC, one can in theory use source coding with side information to reveal, when $l \to \infty$:

$$l^{-1}|M| = I_0 \triangleq H(S_{1...m}(X)|Y) \text{ bits.} \qquad (9.1)$$

When using slices, however, the BCP does not take Y directly into account but instead processes the bits calculated by Claude $S_i(X)$ on one side and the bits calculated by Dominique using the function $E_i(X', S_{1...i-1}(X))$ on the other side. The l bits produced by the slices are, of course, independent for different indexes. Assuming that the chosen BCP is optimal, that is, it reveals only $H(S|E)$ bits for some binary random variables S and E, we obtain the following number of disclosed bits:

$$l^{-1}|M| = I_s \triangleq \sum_{i=1}^{m} H(S_i(X)|E_i(Y, S_{1...i-1}(X))) \geq I_0. \tag{9.2}$$

The inequality follows from the fact that

$$H(S_{1...m}(X)|Y) = \sum_i H(S_i(X)|Y, S_{1...i-1}(X))$$

and

$$H(S_i(X)|E_i(Y, S_{1...i-1}(X))) \geq H(S_i(X)|Y, S_{1...i-1}(X)).$$

The primitive BCP is usually optimized as if E was the result of transmitting S through a binary symmetric channel (BSC–BCP), thus assuming that the bits 0 and 1 have the same probability of occurrence, on both Claude's and Dominique's sides. This is, of course, sub-optimal for unbalanced bits as the actual redundancies cannot be exploited. In this case, the number of bits disclosed becomes:

$$l^{-1}|M| = I_e \triangleq \sum_{i=1}^{m} h(e_i) \geq I_s, \tag{9.3}$$

with $h(e) = -e \log e - (1-e) \log(1-e)$ and

$$e_i = \Pr[S_i(X) \neq E_i(Y, S_{1...i-1}(X))].$$

The inequality follows from Fano's inequality [46] applied to a binary alphabet. In practice, a BSC–BCP is expected to disclose a number of bits that is approximately proportional to $h(e)$, e.g., $(1+\epsilon)h(e)$ for some overhead constant ϵ and thus $l^{-1}|M| \geq I_e$.

Note that in the case of asymptotically large dimensions, $d \to \infty$, the quantities I_0, I_s and I_e tend to the same limit $dH(\Psi(X^{(1)})|Y^{(1)})$ since the first slices can be completely disclosed, determining the remaining ones with arbitrarily small error probabilities, as shown above.

Bit assignments

Among all possible slice functions $S_{1...m}(X)$, there are many equivalent bit assignments. It is a valid assumption that the BCP works equally well with zeroes and ones inverted. Consequently, changing $S_1(X)$ into $S'_1(X) = S_1(X) + 1$ (modulo 2) does not modify the efficiency of SEC. For a slice $S_i(X)$, there are even more equivalent assignments. In general, $S'_i(X) = S_i(X) + f(S_{1...i-1}(X))$ gives an assignment equivalent to $S_i(X)$ for an arbitrary binary function $f(s_1, \ldots, s_{i-1})$. This follows from the knowledge of the previous slices upon evaluation of a slice estimator: The slice estimator E_i only needs to distinguish between zeroes and ones given the value of the previous slices $S_{1...i-1}$. For each slice $i \in \{1...m\}$, there are thus $2^{2^{i-1}}$ equivalent assignments. For the m slices together, this gives

$$\prod_{i \in \{1...m\}} 2^{2^{i-1}} = 2^{(2^m-1)}$$

equivalent assignments.

If we restrict ourselves to slice functions $S_{1...m}(X)$ that define a bijection from $\mathcal{X}$ to $\{0,1\}^m$, there are $2^m!$ such possible functions, which can be grouped into equivalence classes of size $2^{(2^m-1)}$ each. There are thus

$$N_{\text{classes}} = 2^m!/2^{(2^m-1)}$$

such classes.

As an example, let $m = 2$. This gives us $N_{\text{classes}} = 3$ different assignments for a quaternary alphabet, $\mathcal{X} = \{0,1,2,3\}$. Let us denote an assignment by $(S_{12}(0), S_{12}(1), S_{12}(2), S_{12}(3))$. In the first class of equivalence, one can find the binary representation of the numbers X: $(00, 01, 10, 11)$. The second one contains the inverse binary representation, that is, with the least significant bit first: $(00, 10, 01, 11)$. In the third one, one can find $(00, 10, 11, 01)$, which is a Gray code [71, 184] with inverted bits.

For $m = 3$, there are $N_{\text{classes}} = 315$ different assignments, for $m = 4$, $N_{\text{classes}} = 638\,512\,875$, and for $m = 5$, $N_{\text{classes}} \approx 1.1225 \times 10^{26}$. This, of course, grows very quickly with m.

Optimal slice estimators

Maximizing the global efficiency of the sliced error correction protocol for a given pair of variables X and Y is not a simple task because the number of key bits produced and leaked with slice i recursively depends on the design of the previous slices $1...i-1$. For this reason, our goal in this section is simply to minimize $l^{-1}|M|$ by acting on each slice estimator E_i independently. More precisely, we wish to minimize each bit error rate e_i, of which $h(e_i)$ is an increasing function for $0 \leq e_i < 1/2$, so as to locally minimize the number of leaked bits $l^{-1}|M|$ without changing the number of produced bits $H(\Psi(X))$. This approach applies to both perfect and non-perfect BCPs, and results in an explicit expression for $E_i(y, s'_{1...i-1})$; see Eq. (9.4).

The error rate in slice i is the probability that Dominique's slice estimator yields a result different from Claude's slice, and can be expanded as

$$e_i = \sum_y \sum_{s' \in \{0,1\}^{i-1}} \Pr[S_i(X) \neq E_i(y, s') \wedge S_{1...i-1}(X) = s' \wedge Y = y].$$

Each term of the right-hand side of the above equation sums $P_{XY}(x,y)$ over non-overlapping areas of the $\mathcal{X} \times \mathcal{Y}$ set, namely $\{(x,y) : S_{1...i-1}(x) = s'\}$. So, each of these terms can be minimized independently of the others and thus, to minimize e_i, E_i must satisfy:

$$E_i(y, s') = \arg \max_{s \in \{0,1\}} \Pr[S_i(X) = s \mid S_{1...i-1}(X) = s', Y = y], \quad (9.4)$$

with an appropriate tie-breaking rule.

Since the slice estimators are now determined by the slice functions S_i and $P_{XY}(x,y)$, the bit error probability e_i can be evaluated as

$$e_i = \sum_y \sum_{s' \in \{0,1\}^{i-1}} \min_{s \in \{0,1\}} \Pr[S_i(X) = s \wedge S_{1...i-1}(X) = s' \wedge Y = y].$$

Intuitively, the error probability is minimal when the variables y and s' allow to determine $S_i(x)$ without ambiguity.

All that remains is to optimize only the functions S_i, which is done for Gaussian variables in Section 9.3.2.

Binary correction protocols

To be able to use sliced error correction, it is necessary to choose a suitable BCP. There are two trivial protocols that are worth noting. The first one consists in disclosing the slice entirely, while the second does not disclose anything. These are at least of theoretical interest related to the asymptotical optimality of SEC. It is sufficient for Claude to transmit entirely the first $r = \lfloor dH(K(X^{(1)})|Y^{(1)}) + 1 \rfloor$ slices and not to transmit the remaining $m - r$ ones.

Possible BCPs useful in combination with sliced error correction include source coding with side information based on syndromes of error correcting codes, as well as interactive binary error correction protocols.

With SEC, it is not required to use the same protocol for all slices. Depending on the circumstances, one-way and interactive BCPs can be combined. In the particular case of slices with large e_i, for instance, disclosing the entire slice may cost less than interactively correcting it. Overall, the number of bits revealed is:

$$|M| = \sum_i |M_i|, \text{ with } |M_i| = \min(l, f_i(l, e_i)), \quad (9.5)$$

where $f_i(l, e_i)$ is the expected number of bits disclosed by the BCP assigned to slice i working on l bits with a bit error rate equal to e_i.

9.1 Sliced error correction

Complexity

The optimization of slices to maximize $H(S_{1...m}(X)) - l^{-1}|M|$ is a complex problem. More formally, I will show that a decision problem based on the design of slices is NP-complete.

To be independent of a particular BCP, let us consider the maximization of $H(S_{1...m}(X)) - I_e$. The result below easily extends to other measures of disclosed information.

Let us consider the following decision problem. An instance of the slice decision problem is given by the tuple $(m, \mathcal{X}, \mathcal{Y}, P_{XY}, t)$, where $m \in \mathbf{N}$, $m \geq 1$, is the desired number of slices, $(\mathcal{X}, \mathcal{Y}, P_{XY})$ describes the finite discrete variables X and Y, and t is the target. We ask whether there exist slice functions $S_{1...m}(X)$ such that $t = H(S_{1...m}(X)) - I_e$.

Theorem 12 *The slice decision problem described above is NP-complete.*

Proof

We show this by reduction to the subset-sum problem, which is NP-complete [44, 47]. Let $A \subset \mathbf{N}$ be a set of numbers and $a \in \mathbf{N}$ the target. The subset-sum problem asks whether there exists a subset $A' \subseteq A$ whose elements sum to a.

Let us construct an instance of the slice optimization problem corresponding to an instance (A, a) of the subset-sum problem. Let $\mathcal{X} = \mathcal{Y} = A$, $P_{XY}(x, x) = x/\sum_{x \in A} x$ and $P_{XY}(x, y) = 0$ when $y \neq x$. With such a joint probability distribution, X and Y are perfectly correlated and thus $I_e = 0$. Let $m = 1$ and $A' = \{x : S_1(x) = 1\}$. Then, $H(S_1(X)) = h(\Pr[X \in A'])$. Checking whether $a = \sum_{x \in A'} x$ comes down to checking whether $h(\Pr[X \in A']) = h(a/\sum_{x \in A} x)$. However, the function $h(p) = h(1-p)$ is bijective only in the range $[0, 1/2]$, so one has to check both sets A' and $A \setminus A'$.

Clearly the reduction is polynomial in $|A|$ and thus the slice decision problem is NP-complete.

□

The above proof can be trivially extended to other measures of disclosed information, since the reduction uses the special case of X and Y being perfectly correlated. In particular, it also applies to the case of $H(S_{1...m}(X)) - l^{-1}|M|$ with a practical BCP as long as the chosen BCP does not disclose anything ($|M| = 0$) when X and Y are perfectly correlated.

We do not know whether the related optimization problem, namely the maximization of the net secret key rate with sliced error correction, is NP-hard. Yet, Theorem 12 strongly suggests that it is a difficult problem and that we should instead look for approximately optimal solutions.

9.2 Multistage soft decoding

Sliced error correction is asymptotically optimal, that is, its efficiency goes to 1 when the dimension d goes to infinity. When $d = 1$, however, we have $I_e > I_0$ in general, and thus one-dimensional sliced error correction is sub-optimal. A possible way to remove this restriction is to investigate multidimensional slicing, for instance using lattices [137].

As we will see, there is actually another method to increase the reconciliation efficiency while keeping $d = 1$. This method, which is based on multilevel coding and multistage soft decoding (MSD) [33, 177, 188], was recently proposed by Bloch, Thangaraj and McLaughlin [25]. Multilevel coding is a channel coding technique that uses nested constellations. When transposed to reconciliation, this is very similar to slices. So, for our purposes, we keep the same notation as for the SEC. The good efficiency of the method comes from the use of multistage soft decoding combined to efficient LDPC codes.

Let us now describe the encoding and decoding procedures. Claude and Dominique agree on m LDPC codes, one for each slice. Then, Claude sends Dominique the syndromes $H_i S_i(X_{1...l})$, $i = 1 \ldots m$, with H_i the parity check matrix of the associated LDPC code. Upon reception by Dominique, each slice is soft decoded, yielding LLRs

$$L(S_i(X_j)|y_{1...l}) = \ln \frac{\Pr[S_i(X_j) = 0|Y_{1...l} = y_{1...l}]}{\Pr[S_i(X_j) = 1|Y_{1...l} = y_{1...l}]}$$
$$= \ln \frac{\Pr[S_i(X_j) = 0, Y_{1...l} = y_{1...l}]}{\Pr[S_i(X_j) = 1, Y_{1...l} = y_{1...l}]}.$$

The LLR is split into extrinsic and intrinsic information,

$$L(S_i(X_j)|y_{1...l}) = L_{\text{ext}} + L_{\text{int}},$$

in the following way [25]:

$$L_{\text{ext}} = \ln \frac{\Pr[S_i(X_j) = 0|Y_{\setminus j} = y_{\setminus j}]}{\Pr[S_i(X_j) = 1|Y_{\setminus j} = y_{\setminus j}]},$$

$$L_{\text{int}} = \ln \frac{\sum_{s:s_i=0} \Pr[S(X_j) = s, Y_j = y_j] \prod_{i'} \Pr[S_{i'}(X_j) = s_{i'}|Y_{\setminus j} = y_{\setminus j}]}{\sum_{s:s_i=1} \Pr[S(X_j) = s, Y_j = y_j] \prod_{i'} \Pr[S_{i'}(X_j) = s_{i'}|Y_{\setminus j} = y_{\setminus j}]},$$

where $S(X_j) = S_{1...m}(X_j)$, $Y_{\setminus j} = Y_{1...j-1,j+1...l}$ and $y_{\setminus j} = y_{1...j-1,j+1...l}$.

To proceed with MSD, a given slice i is decoded as described in Section 8.5

for several iterations. Then, the resulting extrinsic information is injected as a-priori information for another slice $i' \ne i$, which undergoes several decoding iterations, and so on. The process is iterated several times for the whole set of slices.

Like for turbo codes, the extrinsic information produced by the decoding of one slice is injected as a-priori information into the decoding of other slices. In particular, the value of $\Pr[S_{i'}(X_j) = s_{i'}|Y_{\setminus j} = y_{\setminus j}]$ in the expression of L_{int} is calculated from the extrinsic information coming from the decoding of the other slices.

In theory, the sequential decoding of the slices would be enough. Assuming that the decoding of each slice i requires $H(S_i(X)|S_{1...i-1}(X), Y)$ bits, using the fact that the slices with lower indexes are already decoded, the total number of disclosed bits is

$$\sum_i H(S_i(X)|S_{1...i-1}(X), Y) = H(S_{1...m}(X)|Y) = I_0$$

per key element. This is precisely the fundamental limit below which reconciliation does not work. In practice, however, the soft decoding of slice i can benefit from using not only those with lower indexes $i' < i$ but also all other slices $i' \ne i$.

Note that slices can also be corrected using full disclosure, so as to complement LDPC codes whenever $H(S_i(X)|S_{1...i-1}(X), Y) \approx 1$, that is, when the correlations are poor.

The MSD can be used for the reconciliation of Gaussian key elements as explained in the section below.

9.3 Reconciliation of Gaussian key elements

We must now deal with the reconciliation of information from Gaussian variables $X \sim N(0, \Sigma)$ and $Y = X + \epsilon$, $\epsilon \sim N(0, \sigma)$. In this section, I first give some remarks regarding the reconciliation of continuous key elements. I then describe the practical extraction of a common key from such key elements, both using SEC and MSD.

9.3.1 Remarks about continuous key elements

In the case of continuous variables, I showed in Section 6.6 that, without loss of generality, Claude and Dominique can first convert their variables to discrete ones. However, the problem of continuous reconciliation is not equivalent to known transmission schemes, namely quantization and coded modulation, and therefore deserves special treatment.

In a quantization system, a random input variable X is transmitted over a noiseless discrete channel using the index of the closest code-vector in a given codebook. More precisely, X is encoded as the discrete value $\alpha(X)$, which is then transmitted noiselessly. From this, the decoder decodes $\bar{X} = \beta(\alpha(X))$, and the functions α and β are chosen so as to minimize some average distortion measure d (e.g., the Euclidean distance) between the input and the decoded vector, $E[d(X, \beta(\alpha(X))]$. The codebook design issue has been extensively studied in the literature [72]. With reconciliation, we do not have reproduction vectors, since we are not interested in reproducing the continuous code but rather extracting common discrete information between two random variables X and Y. Furthermore, the quantities to optimize are not the same, namely, the average distortion to minimize for quantization and the amount of secret bits to maximize for this problem. Techniques derived from quantization can be used to find $\Psi(X)$ that maximizes $I(\Psi(X); Y)$, as proposed in [35, 36] and the references therein. Yet, this must still be completed with appropriate reconciliation to extract common information between $\Psi(X)$ and Y.

In a coded modulation system, a binary value is sent over a continuous noisy channel using a vector X belonging to a codebook in an Euclidean space. Trellis-coded modulation and lattice-based coded modulation are instances of this scheme. In this case, the information sent on the channel would be chosen by Claude in a codebook, which is not true for reconciliation.

9.3.2 Constructions

We assume $d = 1$, that is, Claude and Dominique use key elements individually. The idea is to divide the set of real numbers into intervals and to assign bit values to each of these intervals. For SEC, the slice estimators are derived as explained in Section 9.1.2.

For simplicity, we divide the slice design into two smaller independent problems. First, we cut the set of real numbers into a chosen number of intervals – call this process $T(X)$. For the chosen number of intervals, we maximize $I(T(X); Y)$. Second, we assign m binary values to these intervals in such a way that slices can be corrected with as few leaked information as possible.

Optimizing intervals

For both SEC and MSD, one can optimize the intervals by maximizing $I(T(X); Y)$. If the reconciliation is optimal, it produces $H(T(X))$ com-

9.3 Reconciliation of Gaussian key elements

mon bits, discloses I_0 bits and, thus, from Eq. (9.1) gives a net result of $H(T(X)) - H(T(X)|Y) = I(T(X);Y)$ bits. Note that $S_{1...m}(X)$ will be a bijective function of $T(X)$. However, maximizing $I(T(X);Y)$ does not depend on the bit assignment, so as to make the maximization easier.

The process $T(X)$ of dividing the real numbers into $|T|$ intervals is defined by the $|T|-1$ variables $\tau_1 \ldots \tau_{|T|-1}$. The interval t with $1 \leq t \leq |T|$ is then defined by the set $\{x : \tau_{t-1} \leq x < \tau_t\}$ where $\tau_0 = -\infty$ and $\tau_{|T|} = +\infty$.

In [173], the function $I(T(X);Y)$ is numerically maximized under the symmetry constrains $\tau_t = \tau_{|T|-t}$ to reduce the number of variables to process. The results are displayed in Fig. 9.1 below. $I(T(X);Y)$ is bounded from above by $\log|T|$ and goes to $1/2\log(1 + \Sigma^2/\sigma^2)$ as $|T| \to \infty$.

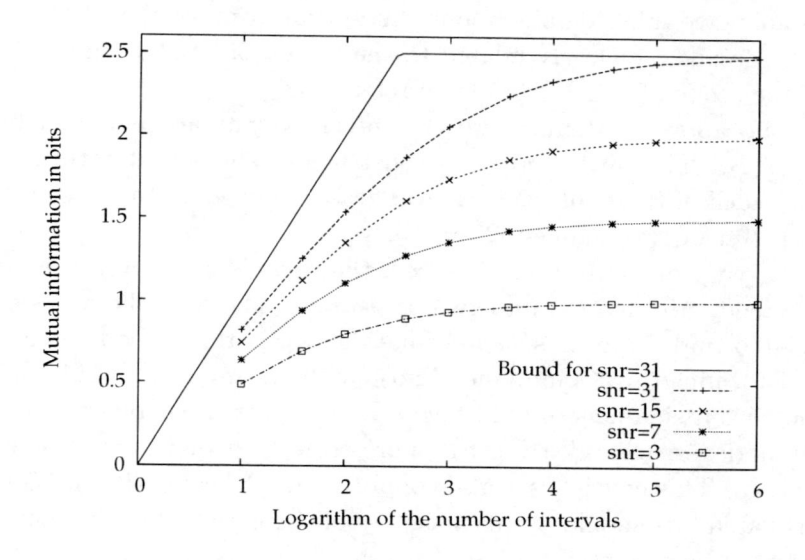

Fig. 9.1. Optimized $I(T(X);Y)$ as a function of $\log|T|$ for various signal-to-noise ratios, with $|T|$ the number of intervals.

For the signal-to-noise ratio $\Sigma^2/\sigma^2 = 15$, an example of division of $\mathcal{X}$ into intervals that maximizes $I(T(X);Y)$ is given in Table 9.1. Note that the generated intervals blend evenly distributed intervals and equal-length intervals. Evenly distributed intervals maximize entropy, whereas equal-length intervals best deal with additive Gaussian noise.

Note that a method for optimizing $T(X)$, including the case of $d > 1$, based on quantization is proposed by Cardinal in [35, 36]. The cells (or intervals if $d = 1$) are optimized iteratively using the Lloyd optimality conditions for vector quantizers [72].

Table 9.1. *Symmetric interval boundaries that maximize* $I(T(X);Y)$, *with* $\Sigma = 1$ *and* $\sigma = 1/\sqrt{3}$.

τ_8	0	$\tau_{12} = -\tau_4$	1.081
$\tau_9 = -\tau_7$	0.254	$\tau_{13} = -\tau_3$	1.411
$\tau_{10} = -\tau_6$	0.514	$\tau_{14} = -\tau_2$	1.808
$\tau_{11} = -\tau_5$	0.768	$\tau_{15} = -\tau_1$	2.347

Reprinted with permission from [173] © 2004 IEEE.

Optimization of sliced error correction

Let me now give more details about the optimization of the SEC. For the MSD, the details are closely related to the theory of LDPC codes, which is beyond the scope of this book; please refer to [25].

From the above procedure, we get intervals that are bounded by the thresholds τ_t. The next step is to construct m slices that return binary values for each of these intervals. Let us restrict ourselves to the case where $|\mathcal{T}|$ is a power of two, namely $|\mathcal{T}| = 2^m$.

In the asymptotic case (i.e., $d \to \infty$), the slices that achieve the asymptotic efficiency are divided into two categories: The first slices are almost uncorrelated and serve as side information, while the second set of slices can be determined by Dominique almost without any error. This suggests that the slice assignment should favor the less correlated bits in the slices with low indexes, while leaving the more correlated bits in the slices with high indexes. This heuristics is also comforted by the concavity of $h(e)$. The non-correlated bits should as much as possible be gathered in the same slice, as $h(e + e') \leq h(e) + h(e')$ for $e + e' \leq 1/2$.

One can also make a comparison with channel coding. Slices with high error rates play the role of sketching a virtual codebook to which Claude's value belongs. After revealing the first few slices, Dominique knows that his/her value lies in a certain number of narrow intervals with wide spaces between them. If Claude had the possibility of choosing a codebook, s/he would pick up a value from a discrete list of values – a situation similar to the one just mentioned except for the interval width. The more slices one chooses, the narrower these codebook-like intervals can be.

The definition of the Gaussian variables X and Y implies that different bits assigned in narrow intervals are less correlated than in wide intervals. I thus propose to use the inverse binary representation of $t - 1$, that is, to assign the least significant bit of the binary representation of $t - 1$ ($0 \leq$

9.3 Reconciliation of Gaussian key elements

$t-1 \leq 2^m - 1$) to the first slice $S_1(x)$ when $\tau_{t-1} \leq x < \tau_t$; then, each bit of $t-1$ is subsequently assigned up to the most significant bit, which is assigned to the last slice $S_m(x)$. More explicitly, $S_i(x) = 1$ if $\tau_{2^i j} \leq x < \tau_{2^i j + 2^{i-1}}$ for $j \in \mathbf{N}$, and $S_i(x) = 0$ otherwise.

Several bit assignment methods have been investigated, based on the equivalence classes of $m = 2$ detailed in Section 9.1.2: the permutations of the bits in the binary representation of $t - 1$ and in its Gray code. The inverse binary representation was found to work best. In [174], I also tried to optimize the bit assignment with simulated annealing. Although less simple assignments were found with this method, they were only slightly better than the inverse binary representation and the general trend of least significant bits first remained.

For the SEC, I thus choose to use the inverse binary representation, which has the additional advantage of being simple to generalize for any m.

Let me now give some numerical examples in the case of a BCP-BSC, as this is the most frequent case in practice. To make the discussion independent of the chosen BCP, we assume it to be perfect and evaluate $H(S_{1...m}(X))$ and $I_e = \sum_i h(e_i)$ for several $(m, \Sigma/\sigma)$ pairs.

We investigate the case of $m = 4$ slices and of signal-to-noise ratio $\Sigma^2/\sigma^2 = 3$. According to Shannon's formula, a maximum of 1 bit can be shared as $I(X;Y) = 1/2 \log(1 + \Sigma^2/\sigma^2) = 1$. Claude's slices follow the definition above, and Dominique's slice estimators are defined as usual using Eq. (9.4). The correction of the first two slices (i.e., the least two significant bits of the interval number) yields error rates that make them almost uncorrelated, namely $e_1 \approx 0.496$ and $e_2 \approx 0.468$. Then come $e_3 \approx 0.25$ and $e_4 \approx 0.02$. So, for $m = 4$, the net amount of information is about $3.78 - 2.95 = 0.83$ bit per key element. Note that the error rate of slice 4 would be much higher (i.e., $e_1 \approx 0.167$) if the slice estimator would not take into account the correction of the first three slices.

Let us investigate another signal-to-noise ratio. When $\Sigma^2/\sigma^2 = 15$, Claude and Dominique can share up to $I(X;Y) = 2$ bits per key element. With $m = 5$, Fig. 9.2 shows a net amount of information of about 1.81 bits per key element and thus an efficiency of $\eta \approx 0.905$.

As expected, the first few error rates (e.g., e_1 and e_2) are high and then the next ones fall dramatically. The first slices are used to narrow down the search among the most likely possibilities Dominique can infer, and then the last slices compose the common information. In Fig. 9.3, these error rates are shown for $m = 4$ when the noise level varies. From the role of sketching a codebook, slices gradually gain the role of extracting information as their error rates decrease with the noise level. In contrast to the asymptotic

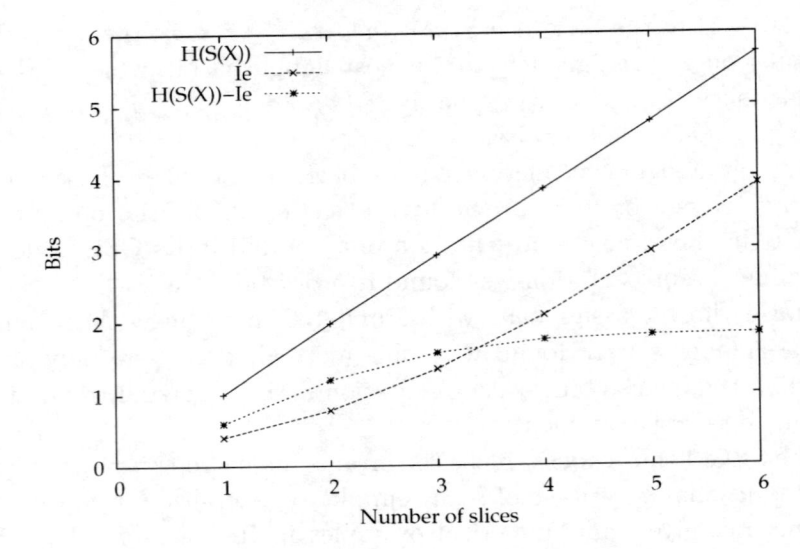

Fig. 9.2. $H(S_{1...m}(X))$, I_e and their difference as a function of the number of slices m when $\Sigma^2/\sigma^2 = 15$.

case, however, there is a smooth transition between slices, giving correction information to slices containing correlated bits.

So far, the optimization of the SEC efficiency was performed by first maximizing $I(T(X); Y)$ as a function of the interval boundaries. With the bit assignment fixed to the inverse binary representation, we can instead try to minimize I_e or more generally to maximize $H(T(X)) - |M|/l$ for a non-perfect BCP.

We first have to model the cost of the BCP. In the case of a perfect BCP, we have $f_i(l, e_i)/l = h(e_i)$. For Cascade, $f_i(l, e_i)/l$ behaves as in Eq. (8.2) and for turbo codes please refer to [136]. Taking this into account and knowing that the slices are defined using the inverse binary representation, we can evaluate $\eta(T)$ given a set of interval boundaries (τ_t) that defines $T(X)$. In [174], the function $\eta(T)$ was maximized with $m = 5$ for various values of $I(X; Y)$. The optimization was done from $I(X; Y) = 0.5$ ($\Sigma^2/\sigma^2 = 1$) to $I(X; Y) = 2.5$ ($\Sigma^2/\sigma^2 = 31$) by steps of 0.1.

9.3.3 Results

Figure 9.4 shows the efficiency of Gaussian key element reconciliation in four different cases. The first curve indicates the efficiency of MSD [25]. The efficiency of SEC when combined with a perfect BCP is plotted in the second

9.3 Reconciliation of Gaussian key elements

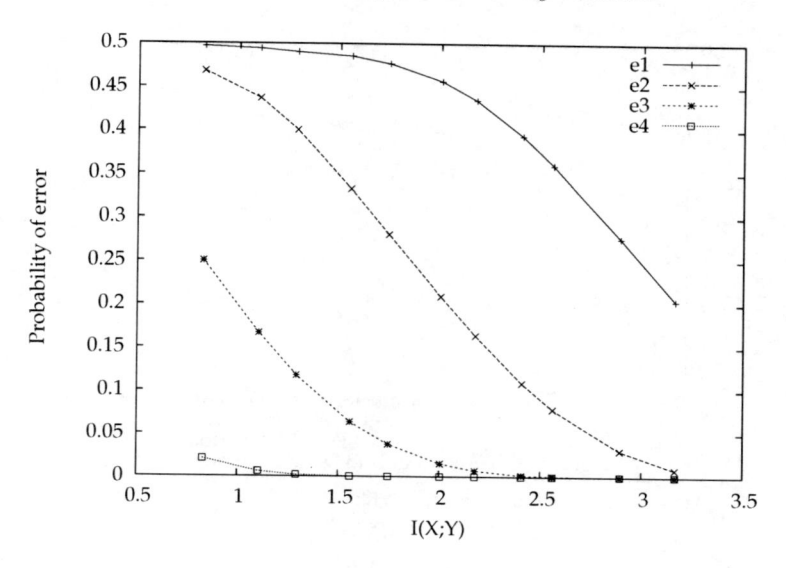

Fig. 9.3. Error rates $e_{1,2,3,4}$ as a function of the correlations between X and Y. Reprinted with permission from [173] © 2004 IEEE.

curve. Then, the third curve shows the case of SEC with Cascade counting only $|M_{\text{ow}}|$, where M_{ow} are the messages sent by Claude to Dominique (and not vice versa) – this assumes that Dominique's parities contain no further information on $\Psi(X)$ and this aspect is discussed in Section 11.5.2. Finally, SEC can be implemented with two different BCPs, each of which performs best in a range of bit error rate. In particular, one can combine Turbo Codes and Cascade and improve the efficiency; the last curve shows the efficiency of SEC with either Turbo Code or Cascade (with all parities counted). Note that to avoid any assumptions, we can instead count in $|M|$ the parities exchanged by both parties – hence the name Double Cascade in the figure.

As one can see in Fig. 9.4, the efficiency drops quickly when $I(X;Y)$ decreases. Correlations between the real numbers X and Y become difficult to exploit. Slices with low error rate, aimed at decreasing $|M|$, cannot be constructed, unless they contain unbalanced bits. Unbalanced bits, on the other hand, do not contribute much to the entropy of the produced key, $H(\Psi(X))$. A part of the inefficiency for low mutual informations comes from the fact that practical BCPs must deal with higher error rates.

The reconciliation of Gaussian key elements using sliced error correction was implemented and applied on experimental data [77]. In this experiment, my colleagues and I used $m = 5$ slices, optimized as explained above. The first two or three slices are too noisy to be processed by a BCP, so it is better

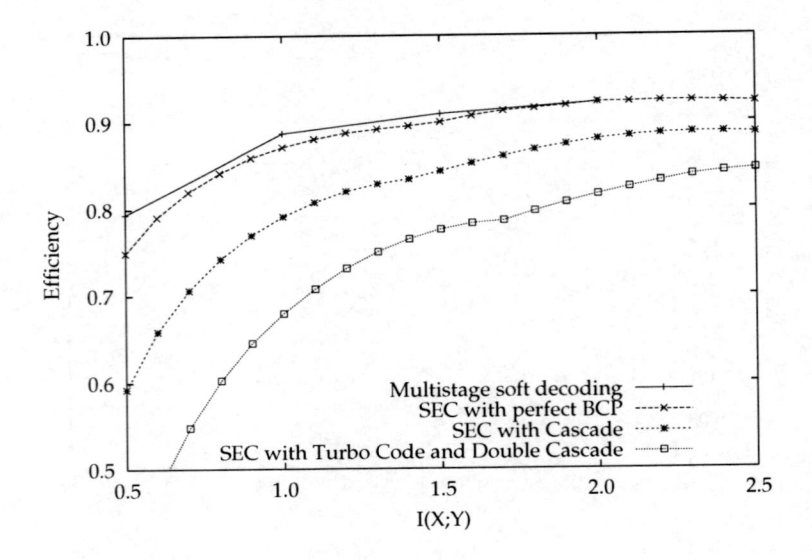

Fig. 9.4. Efficiency η of the reconciliation as a function of $I(X;Y)$.

to disclose such slices entirely. As explained in Section 7.3, the implemented hash function is limited to an input size of up to 110 503 bits. For privacy amplification, fully disclosed slices are not processed with the hash function. We thus used either $l = 36\,800$ or $l = 55\,200$, depending on whether two or three slices were fully disclosed.

The obtained efficiencies are shown in Table 9.2. Note that the evaluation $|M_{\text{ow}}|$ of the number of disclosed bits counts only the bits from Claude to Dominique (one-way) and does not take into account the messages sent back by Dominique during the execution of Cascade. The information gained by Eve on Dominique's parities is calculated otherwise and this aspect will be discussed in Section 11.5.3.

Table 9.2. *Reconciliation of experimental data using Cascade and full disclosure, as in [77]. The evaluation of $|M_{\text{ow}}|$ does not count the messages sent back by Dominique during the execution of Cascade.*

l	$I(X;Y)$	$H(\Psi(X))$	$|M_{\text{ow}}|/l$	η
36 800	2.39	4.63	2.50	89.0%
36 800	2.17	4.48	2.56	88.7%
36 800	1.93	4.33	2.64	87.5%
55 200	1.66	4.70	3.32	83.3%

9.3 Reconciliation of Gaussian key elements

To reduce the interactivity of the reconciliation, the use of turbo codes was studied in [136]. The results are shown in Table 9.3. To take all the disclosed information into account upon reconciliation, the parities coming from both Claude and Dominique are counted in $|M|$, removing any assumptions.

Table 9.3. *Reconciliation of experimental data using Cascade, Turbo Code and full disclosure, as in [136]. The evaluation of $|M|$ takes into account the parities sent back by Dominique during the execution of Cascade.*

l	$I(X;Y)$	$H(\Psi(X))$	$|M|/l$	η
36 800	2.39	4.51	2.51	83.9%
36 800	2.17	4.28	2.49	82.7%
36 800	1.93	4.05	2.49	80.7%
55 200	1.66	4.69	3.40	78.3%

As an example, the details of SEC for the point at $I(X;Y) = 2.17$ are displayed in Table 9.4. The first two slices are entirely disclosed, the third slice is corrected using turbo codes and the last two slices are corrected using Cascade.

Table 9.4. *Detailed figures per slice for the second row of Table 9.3 ($I(X;Y) = 2.17$).*

Slice	e_i	BCP	$|M_i|/l$	$h(e_i)$
1	49.7%	None (full disclosure)	1.00	0.99997
2	34.9%	None (full disclosure)	1.00	0.933
3	6.38%	Turbo code	0.46	0.342
4	0.020%	Cascade	2×0.0052	0.00275
5	6.0×10^{-14}	Cascade	2×0.0040	2.7×10^{-12}

For the MSD, the block length chosen in [25] is $l = 200\,000$; the design of efficient LDPC codes is also described there. Ideally, the number of rows $|M_i|$ of the chosen LDPC code should be as close as possible to $lH(S_i(X)|S_{1...i-1}(X),Y)$. In practice, $|M_i|$ is a little greater to make sure all the errors are corrected.

The efficiency of MSD is displayed in Table 9.5, which increases with $I(X;Y)$. Soft decoding brings a significant improvement in efficiency, as one can see when comparing this table with Table 9.3. The only drawback of MSD is the higher computational complexity: While the SEC allows the correction of each slice only once, MSD requires the decoder to iterate for each slice several times.

Table 9.5. *Simulation results of multistage soft decoding reconciliation with LDPC codes as in [25].*

l	$I(X;Y)$	$H(\Psi(X))$	$|M|/l$	η
200 000	0.5	3.38	2.98	79.4%
200 000	1.0	3.78	2.89	88.7%
200 000	1.5	4.23	2.86	90.9%
200 000	2.0	4.68	2.83	92.2%

As an example, the case of $I(X;Y) = 2$ is expanded in Table 9.6. For the first slices, which bear little correlations between Alice and Bob, MSD uses the full disclosure. The subsequent slices benefit from the disclosed information and are corrected using LDPC codes as described in Sections 9.2 and 8.5.

Table 9.6. *Detailed figures per slice for the last row of Table 9.5 ($I(X;Y) = 2$) [25].*

Slice	BCP	$|M_i|/l$	$H(S_i(X)|S_{1...i-1}(X),Y)$
1	None (full disclosure)	1.00	0.998
2	None (full disclosure)	1.00	0.975
3	LDPC code	0.69	0.668
4	LDPC code	0.14	0.066
5	LDPC code	0.00	0.000

9.4 Conclusion

In this chapter, I described two possible techniques for reconciling the key elements produced by continuous variable QKD protocols, which were then applied to the particular case of Gaussian key elements.

Both discrete and continuous variable QKD protocols, which produce the key elements necessary for secret-key distillation, are examined in the next chapters.

10
The BB84 protocol

The publication of the BB84 protocol by Bennett and Brassard in 1984 [10] marks the beginning of quantum key distribution. Since then, many other protocols have been invented. Yet, BB84 keeps a privileged place in the list of existing protocols: it is the one most analyzed and most often implemented, including those used in commercial products, e.g., [91, 113].

In this chapter, I define the BB84 protocol, although I already described it informally in Section 1.1. The physical implementation is then investigated. Finally, I analyze the eavesdropping strategies against BB84 and deduce the secret key rate.

10.1 Description

Alice chooses binary key elements randomly and independently, denoted by the random variable $X \in \mathcal{X} = \{0, 1\}$. In this protocol, there are two encoding rules, numbered by $i \in \{1, 2\}$. Alice randomly and independently chooses which encoding rule she uses for each key element.

- In case 1, Alice prepares a qubit from the basis $\{|0\rangle, |1\rangle\}$ as

$$X \to |X\rangle.$$

- In case 2, Alice prepares a qubit from the basis $\{|+\rangle, |-\rangle\}$ as

$$X \to 2^{-1/2}(|0\rangle + (-1)^X |1\rangle).$$

On his side, Bob measures either $\mathbf{Z}$ or $\mathbf{X}$, yielding the result $Y_\mathbf{Z}$ or $Y_\mathbf{X}$, choosing at random which observable he measures. After sending a predefined number of qubits, Alice reveals to Bob the encoding rule for each of them. They proceed with sifting, that is, they discard the key elements for which Alice used case 1 (or case 2) and Bob measured $\mathbf{Z}$ (or $\mathbf{X}$). For the remaining (sifted) key elements, we denote Bob's sifted measurements by Y.

From an observer's point of view, the mixed states that Alice sends in case 1 and in case 2 are indistinguishable, i.e.,

$$\frac{1}{2}|0\rangle\langle 0| + \frac{1}{2}|1\rangle\langle 1| = \frac{1}{2}|+\rangle\langle +| + \frac{1}{2}|-\rangle\langle -| = \frac{\mathbf{I}}{2}.$$

As a consequence, Eve cannot obtain any indication as to whether she is measuring a case 1 or case 2 qubit, whatever the statistics she accumulates.

10.2 Implementation of BB84

The implementation of BB84 is a technological challenge. Producing single photons, for instance, is not an easy task. Recent advances, however, show that BB84 can nevertheless be implemented using current technologies. In the following pages, I overview several solutions for implementing BB84. Let me now summarize the different options.

First, the information carriers prescribed by BB84 are ideally *single-photon states*. However, these are difficult to produce, and an alternative solution is to use *weak coherent states*, that is, coherent states with a low average number of photons, to approximate single-photon states. Weak coherent states may sometime contain more than one photon, but the probability of such an event can be controlled. Also, *entangled photon pairs* may be used to produce information carriers.

Second, the photons can either be sent through an *optical fiber* or through the *open air*. This depends on what the application requires. Whereas the optical fiber may be the option of choice for telecommunication networks, the open air solution will obviously be preferred for satellite communications.

Finally, the encoding of the qubit can be done in the *polarization* of the photon or in its *phase*. While phase encoding is usually preferred for photons traveling in an optical fiber, polarization coding is the option of choice for the open air.

10.2.1 Single photon sources vs weak coherent states

In practice, single photons are difficult to produce. Instead, one may turn to weak coherent states as a valid approximation. A coherent state $|\alpha\rangle$ has an average number of photons equal to $\mu = |\alpha|^2/4N_0$. The distribution of the number of photons is Poissonian, that is, the probability of measuring n photons is equal to

$$\Pr[n \text{ photons}] = e^{-\mu}\frac{\mu^n}{n!}.$$

Hence, by creating a coherent state with $\mu \ll 1$, one rarely creates more than one photon. The drawback is that, most of the time, there are no photons at all; $\Pr[n = 0] \approx 1 - \mu$. Given that there is at least one photon, the probability that more than one photon is created can be arbitrarily bounded and behaves as $\Pr[n > 1 | n > 0] \approx \mu/2$. For instance, when $\mu = 0.1$, 90.5% of the pulses are empty. Out of the non-empty pulses, 4.9% of them contain more than one photon.

Coherent states are very easy to create, for instance, using a laser diode. Then, to reduce the average number of photons, an attenuator is used to lower the pulse intensity. In practice, many experiments use an attenuation that reduces the intensity down to $\mu = 0.1$.

The main drawback of using weak coherent states is that the photons in multi-photon pulses all carry the same phase or polarization information. It is, in principle, possible for the eavesdropper to detect a multi-photon pulse. Whenever she detects one, she extracts one photon from the pulse, which she keeps in a quantum memory, and sends the remaining photons to Bob. With such an attack, usually called a photon-number splitting (PNS) attack, the eavesdropper does not disturb the information carried by multi-photon pulses, so she does not increase Bob's error rate, and yet acquires information. Of course, the intensity at Bob's station is lowered, but Eve can, again in principle, use a fiber with lower losses to compensate for the fraction of intensity she gets. The presence of multi-photon pulses must be taken into account as Eve's information in the secret key rate.

Another way to produce photons is to use parametric downconversion. The idea is to create entangled photon pairs. The created photons go in opposite directions, so Alice can detect one of them and let the other one go to Bob. In practice, a photon pair is not created every time Alice triggers the source. However, whenever Alice detects one photon, she knows one was sent to Bob. Because of the entanglement, the photon pair may be modeled as a $|\phi^+\rangle$ state, which carries identical random bits when measured using the same basis:

$$|\phi^+\rangle = 2^{-1/2}(|00\rangle + |11\rangle) = 2^{-1/2}(|++\rangle + |--\rangle).$$

As Alice detects her photon, she measures it using either basis and, thanks to the entanglement, she knows Bob's photon is in the same state she measured. This way of creating photons is thus equivalent to the usual prepare-and-measure procedure.

Note that multiple photon pairs can be created at once, so multi-photon pulses can be sent to Bob. With multiple photon pairs, however, the various pairs are independent of one another and thus cannot be used by Eve to extract information. They may increase the bit error rate at Bob's station, however.

Finally, real single-photon sources are becoming within reach of current technologies. Various approaches are being explored. In particular, two-level quantum systems can only emit one photon when the state makes a transition from the higher-energy level to the lower-energy level. Such systems may be trapped atoms or ions, dye molecules in solvent or nitrogen-vacancy color centers in diamonds. Other approaches involve single electrons in a p-n junction or electron-hole pairs in a semiconductor quantum dot. For a review, please refer to [151].

For quantum key distribution, the advantage of single-photon sources over weak coherent pulses is the longer achievable distances. Since no multi-photon pulses are produced, the information carried by single photons has better secrecy than weak coherent states. Also, BB84 using single-photon sources provides better robustness against dark counts than using weak-coherent states – see Section 10.3.3.

For instance, Alléaume, Beveratos and coworkers implemented a single-photon source using a single nitrogen-vacancy color center in a diamond nanocrystal [2, 19]. They used their photon source to implement polarization-based BB84 in the open air between two wings of a building in Orsay, France. A similar experiment was conducted by Waks and coworkers using a laser-excited single quantum dot in a micropost cavity, which they called a turnstile [178]. The single photons produced are polarized and sent through a meter of air and through a variable attenuator to simulate losses.

10.2.2 Polarization encoding

The polarization of light is a characteristic of its propagation. For a classical plane wave, it is the $\bar{\epsilon}$ vector of Eq. (4.2), which gives a preferred direction to the electrical field in a plane perpendicular to its propagation direction. For a single photon, quantum mechanics describes polarization in a two-dimensional Hilbert space. For the BB84 protocol, we will only use four different polarization directions: the vertical, horizontal, diagonal and anti-diagonal directions. Because the Hilbert space has only two dimensions, the diagonal and anti-diagonal polarization states are actually superpositions of the vertical and horizontal polarization states. More explicitly, we can map polarization states to the following formal quantum states of BB84:

$$|\rightarrow\rangle = |0\rangle,$$
$$|\uparrow\rangle = |1\rangle,$$
$$|\nearrow\rangle = (|\rightarrow\rangle + |\uparrow\rangle)/\sqrt{2} = |+\rangle,$$
$$|\searrow\rangle = (|\rightarrow\rangle - |\uparrow\rangle)/\sqrt{2} = |-\rangle.$$

Polarized light can be created with laser diodes. One option is to have four different lasers, each emitting in one of the four prescribed polarization directions. For each time slot, only one laser fires a short pulse. The pulses of all the lasers are combined using a set of beam splitters to obtain a single source. Another option is to have a single laser diode, whose output polarization is modulated using an active polarization modulator, e.g., a Pockels cell.

On his side, Bob uses a balanced beam splitter to guide the pulse in either of two sets of detectors, one for each basis – see Fig. 10.1. The choice of the basis $\{|0\rangle, |1\rangle\}$ or $\{|+\rangle, |-\rangle\}$ is made passively by the beam splitter. The arm of the diagonal basis uses a half waveplate in one of the arms to rotate the polarization by 45°. For each basis, the light pulse is fed into a polarizing beam splitter, which splits the pulse according to its polarization. Detectors on each arm perform the measurement and output either 0 or 1. The combination of the polarizing beam splitter and of the two photon detectors is thus a way to measure the polarization of the photon.

The polarized pulses produced by Alice can be injected into an optical fiber. It is essential that the fiber preserves the polarization of the pulses, otherwise Bob will not be able to decode the bits sent by Alice properly. Of course, even if the polarization is preserved through the entire trajectory, the fiber itself is not a straight line nor is it planar. As the light travels through the fiber, the polarization can be rotated due to its geometry. Bob can compensate for this effect by rotating the polarization of the incoming photons so that they align on Alice's polarization. This compensation must be active, as the geometry of the fiber may vary over time due to temperature changes.

A variety of other polarization effects can happen in the fiber optics. For instance, birefringence induces two different phase velocities for two orthogonal polarization states, polarization mode dispersion causes orthogonal polarization states to have different group velocities, and there may be polarization-dependent losses [64].

All in all, the difficulty of preserving polarization over optical fibers makes this encoding less suited for quantum key distribution in telecommunica-

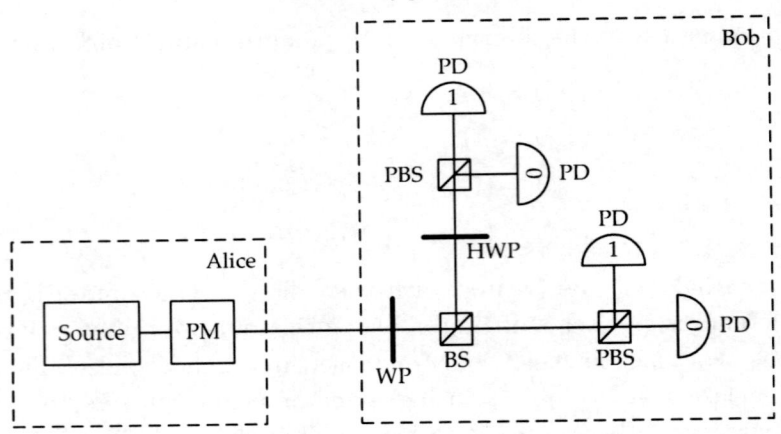

Fig. 10.1. Schematic implementation of BB84 using polarization states. Alice's side comprises a photon source (Source) and a polarization modulator (PM), although she could as well combine the output of four different sources, each with a different polarization. As the photon enters Bob's station, it goes first inside a waveplate (WP), which corrects polarization changes due to the fiber. The beam splitter (BS) passively branches the photon to one of the two possible measurement bases. One of the outputs goes inside a half waveplate (HWP) to rotate the polarization by 45°. The polarizing beam splitters (PBS) select the photons based on their polarization state. The photon detectors (PD) are associated with either the value "0" or with the value "1".

tion networks. Yet some successful experiments have been conducted. For instance, Muller, Zbinden and Gisin implemented polarization-based BB84 over a 23 km long fiber [128]. This fiber was part of a cable used by Swisscom for phone conversation, which ran between Geneva and Nyon, Switzerland, under Lake Geneva.

Polarization-based implementation is actually better suited for open air quantum key distribution. Notably, the very first experimental implementation of BB84 by Bennett and coworkers actually used polarization of light over 30 cm of air [13].

The polarization of a photon traveling in the open air is better preserved than in an optical fiber. However, the major difficulty of open air quantum key distribution is the background light, especially during the day, but also during the night because of the moon and the city lights. To distinguish a single photon from the environment, narrow spectral, spatial and temporal filtering must be used. Detectors are active only during a short time window around the expected arrival time. Only photons having the proper wavelength and incoming angle are allowed to enter the detector.

Since 1992, several experiments have been performed in the open air.

10.2 Implementation of BB84

Among the longest distances achieved, a key distribution over 10 km was proposed by Hughes and coworkers [87]. The experiment took place in New Mexico, USA, both in daylight and at night. Another key distribution was performed in the mountains of southern Germany by Kurtsiefer and coworkers [105]. The distance between Alice's and Bob's stations was more than 23 km.

10.2.3 Phase encoding

The phase encoding is the most popular approach for implementing BB84 in an optical fiber. It is based on the Mach–Zehnder interferometer, which splits a single photon into two "half"-photons, each traveling along a different interference path, and makes both "halves" interfere.

Mach–Zehnder interferometer

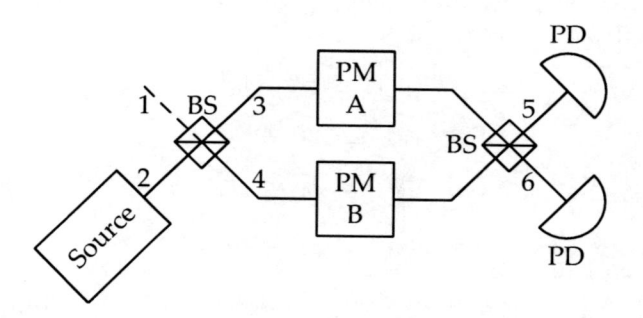

Fig. 10.2. Mach–Zehnder interferometer. The source (Source) is input in the arm 2 of the first beam splitter (BS), while the arm 1 inputs vacuum. Output branches 3 and 4 undergo a phase shift of ϕ_A and ϕ_B, respectively, in the phase modulators (PMA and PMB). The branches are combined in a second beam splitter (BS), whose output arms 5 and 6 each enters a photon detector (PD).

Let us consider the experiment of Fig. 10.2. A single photon is input into the first beam splitter. The input state is $|01\rangle_{n_1 n_2}$ in the two-mode Fock basis, i.e., there are no photons in n_1 and a single photon in n_2. For a balanced beam splitter, the state is transformed as

$$|01\rangle_{n_1 n_2} \to (|10\rangle_{n_3 n_4} + i|01\rangle_{n_3 n_4})/\sqrt{2}.$$

After the beam splitter, half of the probability goes into each of the two arms. The reflected part undergoes a $\pi/2$ phase shift, hence the i factor for the photon in n_4. Then, a phase shift occurs in the two arms. Only the relative phase $\phi = \phi_A - \phi_B$ matters, and the state at the input of the second beam splitter can be written as

$$(e^{i\phi/2}|10\rangle_{n_3 n_4} + ie^{-i\phi/2}|01\rangle_{n_3 n_4})/\sqrt{2}.$$

For the second beam splitter, the reasoning is the same. The photon in n_3 is transformed as $|10\rangle_{n_3 n_4} \to (i|10\rangle_{n_5 n_6} + |01\rangle_{n_5 n_6})/\sqrt{2}$, while the photon in n_4 is transformed as $|01\rangle_{n_3 n_4} \to (|10\rangle_{n_5 n_6} + i|01\rangle_{n_5 n_6})/\sqrt{2}$. Depending on the phase shift $\phi = 0, \pi/2, \pi$ or $3\pi/4$, a little bit of algebra yields the following output states:

$$|\psi(\phi = 0)\rangle = i|01\rangle_{n_5 n_6},$$
$$|\psi(\phi = \pi/2)\rangle = (i|01\rangle_{n_5 n_6} + i|10\rangle_{n_5 n_6})/\sqrt{2},$$
$$|\psi(\phi = \pi)\rangle = i|10\rangle_{n_5 n_6},$$
$$|\psi(\phi = 3\pi/2)\rangle = (i|01\rangle_{n_5 n_6} - i|10\rangle_{n_5 n_6})/\sqrt{2}.$$

The states at the second beam splitter are thus formally equivalent to the four BB84 states,

$$|\psi(\phi = 0)\rangle = |0\rangle,$$
$$|\psi(\phi = \frac{\pi}{2})\rangle = |+\rangle,$$
$$|\psi(\phi = \pi)\rangle = |1\rangle,$$
$$|\psi(\phi = 3\frac{\pi}{2})\rangle = |-\rangle.$$

Alice controls $\phi_A \in \{0, \pi/2, \pi, 3\pi/2\}$ to select one of the four states. Bob always measures the incoming state in the $\{|0\rangle, |1\rangle\}$ basis, although he can choose the value of ϕ_B in $\{0, \pi/2\}$ to simulate the selection of basis. Conclusive measurement occurs when $\phi_B = 0 \wedge \phi_A \in \{0, \pi\}$ and when $\phi_B = \pi/2 \wedge \phi_A \in \{\pi/2, 3\pi/2\}$.

Double Mach–Zehnder interferometer

Phase encoding as depicted in Fig. 10.2 is difficult because the length of the two arms of the fiber optics must match very precisely, within a fraction of the wavelength. Assuming tens of kilometers between Alice and Bob, any change of temperature will dilate or shrink the fiber by a few orders of magnitude longer than the wavelength.

A way to prevent this is to use the double Mach–Zehnder construction, as depicted in Fig. 10.3. The interferometers are unbalanced, that is, their arms do not have equal lengths. The photon emitted by Alice can travel either in the two long arms, in the two short ones or in a short and in a long arm. The

10.2 Implementation of BB84

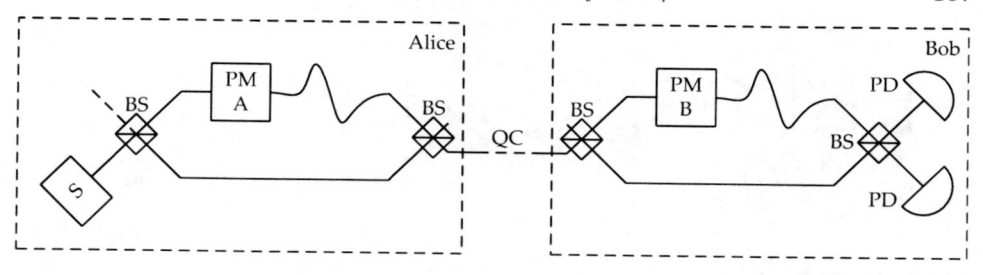

Fig. 10.3. Double Mach–Zehnder interferometer. Alice's station comprises a photon source (S), a first beam splitter (BS), a phase modulator (PMA) and a second beam splitter (BS). Notice that the upper branch is longer than the lower one. The signals are recombined and sent through the quantum channel (QC). Bob's station is similar to Alice's, with a different phase modulator (PMB). The long and short branches are combined using the fourth beam splitter (BS). The output arms are connected to photon detectors (PD).

arrival time will be different and thus these cases can be distinguished. Note that the arrival time will be the same for a photon traveling first in the long arm then in the short arm, or vice-versa. Hence, if we look only at the middle arrival time, the two "half"-photons interfere, one undergoing Alice's phase modulator and one undergoing Bob's. By using this trick, the two "half"-photons travel in the same quantum channel, hence a large portion of the fiber may have length variations, which do not influence the interferences. Only the parts corresponding to the unbalanced interferometers must have their lengths carefully controlled or compensated for.

Plug-and-play construction

A further construction, called *plug-and-play*, allows for an automatic optical alignment and is illustrated in Fig. 10.4. The idea is to combine the double Mach–Zehnder construction with time multiplexing and orthogonal polarization. Let me describe this construction in more detail.

In this construction, the pulses are initiated by Bob, not Alice. Yet, the pulses sent by Bob are always identical and do not contain any information. Bob produces a (classical) strong pulse, which is split into two pulses by the beam splitter BS2. Half of the intensity goes into the short arm (pulse p_1), while the other half goes into the long arm (pulse p_2). At this point, Bob's phase modulator is inactive, so the long arm does nothing but delay the pulse. The polarization of p_2 is chosen so that it is transmitted by the polarizing beam splitter into the quantum channel. In the short arm, however, the polarization of the pulse p_1 is rotated by 90° so that it is

168   *The BB84 protocol*

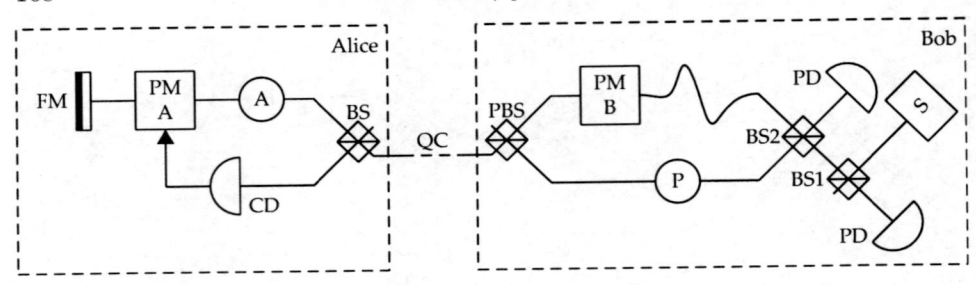

Fig. 10.4. Plug-and-play construction. Bob's station comprises a source (S), which is injected into the fiber through a beam splitter (BS1). The two arms of Bob's station are connected by a beam splitter (BS2) and a polarizing beam splitter (PBS). The long arm contains a phase modulator (PMB), which is active only when the pulses come back to Bob's station. The short arm contains a polarizer (P) that rotates the polarization by 90°. The pulses travel through the quantum channel (QC). Alice's station comprises a beam splitter (BS). On the upper arm, the pulses are directed through an attenuator (A) and a phase modulator (PMA), which is active only for the second pulse. The Faraday mirror (FM) reflects the pulses and rotates their polarization by 90°. On the lower arm, the pulses are detected by a classical detector (CD).

reflected by the polarizing beam splitter and injected into the quantum channel.

At the output of Bob's station, the two pulses are transmitted at slightly different times (first p_1, then p_2) and with orthogonal polarizations.

When the first pulse (p_1) enters Alice's station, half of the intensity goes into the upper arm. At this point, Alice's phase modulator is also inactive. The pulse is attenuated by an attenuator and reflected by a Faraday mirror, which rotates its polarization by 90°.

Half of p_1 also enters the lower arm and the detector. The detector triggers Alice's phase modulator, which is now ready for the second pulse. When the second pulse p_2 enters Alice's station, half of the intensity goes into the upper arm. As for the first pulse, the second pulse is attenuated and its polarization is rotated by 90°. Since Alice's phase modulator is now active, the phase of p_2 is modulated by ϕ_A.

The pulses are now sent back to Bob. At the output of Alice's station, the two pulses have been strongly attenuated and have a quantum nature. They are still separated by the same delay and still have orthogonal polarizations; the second pulse, p_2, however, is phase-modulated and encodes Alice's secret key bit.

Upon entering Bob's station, p_1 is transmitted by the polarizing beam splitter. This is due to the Faraday mirror, which rotated the pulses. The pulse p_1 now goes through the long arm and gets phase-modulated by ϕ_B. Likewise, the second pulse p_2 is reflected by the polarizing beam splitter and goes through the short arm. Because they traveled the same distance, the two pulses arrive at the same time at the beam splitter and interact as in the regular Mach–Zehnder interferometer.

This construction has the advantage of providing self-compensation for variations in the circuit length and for polarization changes due to the fiber. First, the round-trip length is exactly the same for the two pulses. They both go through the short and the long arms of Bob's station, only at different times. Second, the use of the Faraday mirror cancels the polarization changes on the fiber. Without going into details, all the polarization changes, including birefringence, when the pulses are traveling from Bob to Alice, are compensated for during the return path. The polarization of a pulse coming back to Bob's polarizing beam splitter is exactly orthogonal to the polarization it had when leaving Bob's station.

This section concludes the overview of phase encoding techniques.

The experimental implementations using phase encoding are numerous. There is the plug-and-play implementation by Zbinden, Ribordy, Stucki and coworkers [149, 171, 191], which was used to distribute a key between Geneva and Nyon, Switzerland, under Lake Geneva (the same fiber as in [128]) and over 67 km between Geneva and Lausanne, Switzerland. Other systems based on phase encoding are proposed by Bethune, Navarro and Risk [18] and by Bourennane and coworkers [26]. Finally, Gobby, Yuan and Shields demonstrated a working setup with a range of up to 122 km [66].

10.2.4 Photon detectors

So far, we have mainly talked about how Alice generates information carriers and how she does the encoding. On Bob's side, he must use a *photon detector* to measure the polarization or the phase of the incoming photons.

Photon detectors for quantum key distribution are usually avalanche photodiodes. An electrical voltage is applied to a semiconductor, such as silicon (Si), germanium (Ge) or indium gallium arsenide (InGaAs). The applied voltage is high enough (i.e., above the breakdown threshold) so that, when a photon hits the semiconductor, it is absorbed and causes an avalanche of electrons. This avalanche of electrons creates an electrical current, and hence an electrical signal.

Once a photon is detected, the detector must stop the avalanche, so that it is ready for another photon to be detected. This can be done by lowering the voltage below the breakdown threshold. One way of doing this is to use a resistor connected in series with the photodiode. When the avalanche starts, the current passing through the resistor causes the voltage to decrease. Alternatively, an active quenching circuit can be applied, so as to actively lower the voltage when an avalanche is detected. A third solution is to apply the voltage only during short time windows, when a photon is expected to arrive. For instance, if Alice and Bob have a synchronized clock, they can activate the photodiode periodically.

Photon detectors are not perfect, and two main kinds of noise events can occur. First, some incoming photons may be absorbed without causing an avalanche. The fraction of detected photons is called the *efficiency* of the detector, which should be as high as possible. Second, avalanches can occur even if no photon arrives. This may be caused by thermal noise or by band-to-band tunneling processes. The event of an avalanche without an incoming photon is called a *dark count*.

Also, photon detectors may not operate arbitrarily quickly. After an avalanche occurs, some trapping levels of the semiconductor get populated with electrons, which in turn may later cause an avalanche without an incoming photon, called an *afterpulse*. After a photon has been detected, the detector must wait some fixed recovery time so that the trapping levels are emptied. Afterwards, the high voltage may be applied again. So, to avoid afterpulses, the photon detectors impose a maximal photon detection frequency.

10.3 Eavesdropping and secret key rate

In this section, the BB84 quantum key distribution protocol is connected to secret-key distillation. In particular, there are two quantities that we need to evaluate: $I(X;Y)$, the information shared by the legitimate parties, and $I(X;Z)$, the amount of information that the eavesdropper was able to get. We focus on *individual* eavesdropping strategies. For other eavesdropping strategies, the analysis is more involved and will be covered in Chapter 12.

For an individual strategy, Eve sits between Alice and Bob and intercepts or probes one qubit at a time. She can perform whatever operation that is allowed by quantum mechanics, but she cannot make two (or more) consecutive qubits interact so as to possibly get more information. The important point here is that, with such assumptions, we can use the repetitive results of Section 6.4. The secret key rate that the legitimate parties can obtain with perfect reconciliation techniques is the difference of such information rates:

$$S = \max\{I(X;Y) - I(X;Z), I(X;Y) - I(Y;Z)\}.$$

Also, we first assume that Alice and Bob send single-photon states. The case of weak coherent states is covered in Section 10.3.4.

Note that BB84 is fairly symmetric in Alice and Bob. During reconciliation, they exchange parities to find where their bit differs and they correct such differences. Since these differences are known to the eavesdropper – reconciliation is performed over a public channel – it does not matter whether Alice's or Bob's bits serve as a key. Hence, without loss of generality, we assume that Eve tries to guess Alice's bits and that she tries to maximize $I(X;Z)$.

10.3.1 Intercept and resend

The first eavesdropping strategy that we may think of is the simple intercept and resend strategy. In this case, Eve sits between Alice and Bob and measures the qubits as they arrive. Of course, Eve does not know in advance in which basis Alice encoded the secret key bit. Like Bob, she may try at random to measure either in the $\{|0\rangle, |1\rangle\}$ or the $\{|+\rangle, |-\rangle\}$ basis. She may or may not get a significant result, either with a 50% probability. After her measurement, she has to send something to Bob. If she does not, Bob will not take this photon into account, considering it as being lost due to the attenuation in the fiber; this would make her eavesdropping attempt useless. The best she can do is to send a qubit compatible with her measurement, so she sends the result of her measurement in her measurement basis.

To analyze this strategy quantitatively, we may restrict our analysis by assuming that Eve makes a measurement in the $\{|0\rangle, |1\rangle\}$ basis. Because of the symmetry in the bases, the same results apply when she uses the other basis.

When Alice sends either $|\psi\rangle = |0\rangle$ or $|\psi\rangle = |1\rangle$, Eve gets the key bit without any error since her measurement is compatible with Alice's choice. When she resends the state she measured, Bob does not notice any difference, as he gets exactly the state Alice intended to send. Furthermore, when the bases are disclosed, Eve gets the confirmation that she measured in the good basis. In this case, she acquires one bit of information and causes no error,

$$I(X;Z||\psi\rangle \in \{|0\rangle, |1\rangle\}) = 1 \text{ and } \Pr[X \neq Y||\psi\rangle \in \{|0\rangle, |1\rangle\}] = 0.$$

When Alice sends either $|\psi\rangle = |+\rangle$ or $|\psi\rangle = |-\rangle$, Eve gets an uncorrelated result as she measures in the wrong basis. Bob, on his side, also gets uncorrelated results, as Eve resends the wrong state, $|+\rangle$ or $|-\rangle$, and he thus gets an error about half of the time. In this case, we have

$$I(X;Z||\psi\rangle \in \{|+\rangle, |-\rangle\}) = 0 \text{ and } \Pr[X \neq Y||\psi\rangle \in \{|+\rangle, |-\rangle\}] = 1/2.$$

Overall, the intercept and resend strategy gives Eve $I(X;Z) = 1/2$ of acquired information and causes an error rate $e = \Pr[X \neq Y] = 1/4$ on Alice's and Bob's bits. Of course, Eve may apply her strategy only a fraction p_{IR} of the time so as to make herself less noticeable. In this case, she induces an error rate $e = p_{\text{IR}}/4$ and obtains $I(X;Z) = p_{\text{IR}}/2 = 2e$ bits of information per photon. In all cases, the information shared by Alice and Bob is $I(X;Y) = 1 - h(e)$.

10.3.2 Optimal individual eavesdropping strategy

The intercept and resend strategy is fairly brutal in the sense that Eve first converts quantum information (i.e., the qubit sent by Alice) into classical information by performing a measurement, and then she tries to recreate quantum information to send a qubit to Bob. Quantum information is fairly fragile and Eve would probably cause fewer perturbations by making sure the photon sent to Bob remains in the quantum realm during the whole process. In fact, she could make Alice's photon interact with a probe, which she would then measure so as to get information correlated to Alice's secret key bit. Furthermore, she could hold her probe untouched until Alice and Bob perform sifting, during which they reveal the basis used to encode a particular qubit. This way, Eve would be able to best choose her measurement as a function of the encoding basis.

It is not a trivial question to determine Eve's optimal strategy for all the possible operations that Eve could perform, which are compatible with quantum mechanics. Nevertheless, this problem was solved by Fuchs and coworkers [61]. We will not detail their result further, but we instead present an equivalent one that shares interesting similarity with quantum cloning machines.

We saw in Section 4.5 that perfectly cloning an unknown state is impossible by the laws of quantum mechanics. Imagine for a second that it would be possible. Eve could simply clone the photon traveling between Alice and Bob, hold it into a quantum memory (e.g., a fiber loop) until they reveal which basis they used, and then make a measurement. She would know the secret key bit exactly, while Bob would not notice any difference. This would contradict the point of quantum cryptography.

Nevertheless, we saw in Section 4.5 that the *imperfect* cloning of a quantum state is possible. Eve can do the following: she clones the photon sent by Alice, sends one of the clones to Bob, holds her clone in a quantum memory until the bases are revealed and measures it in the appropriate basis. Since perfect cloning is not possible, there is a trade-off: she either makes

10.3 Eavesdropping and secret key rate

an accurate clone for her, in which case the clone sent to Bob is poor and is likely to induce an error. Or she makes an accurate clone for Bob, so as to go unnoticed, but then she does not get much information on the key.

Note that Eve is not interested in the cloning of any unknown quantum state; she is, instead, interested in the cloning of the particular family of states used in BB84. Actually, it can be shown [41] that the best cloning machine for the BB84 states turns out to be the same as the best cloning machine for all the states of the form $\cos\omega|0\rangle + \sin\omega|1\rangle$ for $0 \leq \omega < 2\pi$; this is the so-called *phase-covariant cloning machine*.

It is interesting to note that the best phase-covariant cloning machine yields the best eavesdropping strategy, as found by Bruß and coworkers [30]. In the context of quantum cloning, the optimality is on the fidelity of the clones with regard to the input state. Nevertheless, the optimality of the fidelity coincides with the optimality of the eavesdropping strategy in terms of acquired information for a given induced error rate.

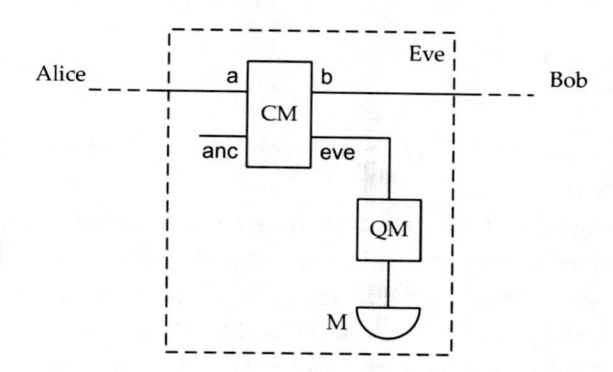

Fig. 10.5. Eavesdropping using a cloning machine. Eve installs a (phase-covariant) cloning machine (CM) on the quantum channel between Alice and Bob. The input **a** comes from Alice, while **anc** is an ancilla. The output clone in **b** is sent to Bob, while the one in **eve** stays in a quantum memory (QM) until the bases are revealed. Eve reads out the information out of her measurement device (M).

Let us describe the best phase-covariant cloning machine. For this, let us denote the eigenstates of the Pauli matrix **Y** by $|0_y\rangle = 2^{-1/2}(|0\rangle + i|1\rangle)$ and $|1_y\rangle = 2^{-1/2}(|0\rangle - i|1\rangle)$. As depicted in Fig. 10.5, the cloning machine takes as its inputs the state to clone (i.e., the photon sent by Alice) in **a** and a fixed ancilla state $|0_y\rangle_{\text{anc}}$. The output of the machine consists of two qubits, one in the system **b** that is sent to Bob and one in the system **eve** that the eavesdropper keeps. The cloning machine is described as follows,

$$|0_y\rangle_a |0_y\rangle_{anc} \to |0_y\rangle_b |0_y\rangle_{eve}$$
$$|1_y\rangle_a |0_y\rangle_{anc} \to \cos\phi |1_y\rangle_b |0_y\rangle_{eve} + \sin\phi |0_y\rangle_b |1_y\rangle_{eve},$$

with $0 \leq \phi \leq \pi/2$.

The parameter ϕ indicates the trade-off between the quality of the two clones. When $\phi = 0$, the state in a is transferred into b without any change, while the eavesdropper gets nothing more than the fixed ancilla. And vice versa when $\phi = \pi/2$: the state in a is transferred to the eavesdropper eve, while b gets the fixed ancilla state. For $\phi = \pi/4$, the two clones have identical quality. For $\phi < \pi/4$, the clone in b gets a better quality than the one in eve, and vice versa for $\phi > \pi/4$.

The cloning machine, expressed in the more familiar basis $\{|0\rangle, |1\rangle\}$, has the following form:

$$|a\rangle_a |0_y\rangle_{anc} \to \sum_{b,c \in \{0,1\}} \alpha_{abc} |b\rangle_b |c\rangle_{eve}, \quad a \in \{0,1\},$$

with

$$\alpha_{abc} = e^{(a+b+c)\pi/2}((-1)^a + (-1)^b \cos\phi + (-1)^c \sin\phi).$$

Let us now analyze the amount of information that Eve gains using this strategy and the error rate she induces. Since this cloning machine clones all of BB84's four states equally well, we can assume without loss of generality that Alice sends the state $|0\rangle$ – the result will be the same for the other states. The error rate between Alice's and Bob's key elements can be obtained by calculating the probability that Bob measures $|1\rangle$ although Alice sent $|0\rangle$. This can be obtained by calculating the probability mass on the $|1\rangle_b$ state, namely

$$e = \sum_c |\alpha_{01c}|^2 = \frac{1 - \cos\phi}{2},$$

and thus

$$I(X;Y) = 1 - h\left(\frac{1 - \cos\phi}{2}\right).$$

On her side, Eve tries to infer the correct bit value by measuring her part eve of the state. Still assuming that Alice sent $|0\rangle$, the probability of Eve guessing wrong is the probability mass on the $|1\rangle_{\text{eve}}$ state, that is,

$$\sum_b |\alpha_{0b1}|^2 = \frac{1-\sin\phi}{2} \text{ and} \tag{10.1}$$

$$I(X;Z) = 1 - h\left(\frac{1-\sin\phi}{2}\right) = 1 - h\left(\frac{1}{2} - \sqrt{e(1-e)}\right). \tag{10.2}$$

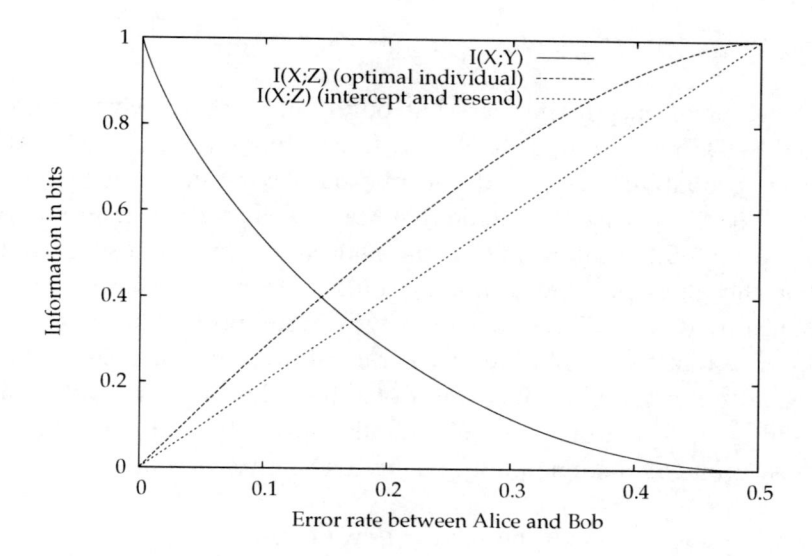

Fig. 10.6. Information shared by the legitimate parties ($I(X;Y)$) or gained by an eavesdropper ($I(X;Z)$) as a function of the error rate e. Both the simple intercept and resend strategy and the optimal individual strategy are shown.

These quantities are plotted in Fig. 10.6. When $e = (1-2^{-1/2})/2 \approx 14.6\%$, Eve has as much information as Alice and Bob. Above this error rate, quantum key distribution cannot work using one-way secret-key distillation and the legitimate parties need to abort the protocol or move on to the next block.

Note that the intercept and resend strategy is also depicted in Fig. 10.6. Clearly, the information gained by Eve is lower than for the phase-covariant cloning machine.

10.3.3 Practical sifting and error rate

In practice, errors are not often caused by an eavesdropper but rather by imperfections in the setup. Let us calculate the sifted key rate and the error rate due to different causes that occur in practice.

The sifted key rate is the number of bits per second that Bob receives, excluding those that come from incompatible measurements. Alice fires pulses at a given repetition rate R_{rep}, whether she sends weak coherent states or single photons.

Assuming that she sends weak coherent states with μ photons on average, the sifted key rate for the legitimate photons traveling from Alice to Bob reads

$$R_{\text{AB}} = R_{\text{rep}} \frac{1}{2} p_{\text{TW}} (1 - e^{-\mu}) T \eta_{\text{PD}}.$$

The 1/2 factor is due to the fact that only half of the received pulses are measured by Bob in a compatible basis. Depending on the setup, there may be a given probability p_{TW} that the photon arrives in the expected time window. This is the case for the double Mach–Zehnder construction, where there is a probability $p_{\text{TW}} = 1/2$ for the photon to travel in a short and long arm. For the plug-and-play setup, the pulses always interfere at the same time, hence $p_{\text{TW}} = 1$. The factor $(1 - e^{-\mu})$ is the probability that at least one photon is emitted by the weak coherent state. The variable T indicates the attenuation of the fiber, i.e., the probability that a given photon arrives at the end of the quantum channel. Finally, η_{PD} is the detector efficiency.

For a setup using a single-photon source, this rate becomes

$$R_{\text{AB}} = R_{\text{rep}} \frac{1}{2} p_{\text{TW}} T \eta_{\text{PD}}.$$

The sifted key rate is composed of both the legitimate photons sent by Alice and the dark counts. For each time slot, Bob expects a photon, but a dark count may as well happen at the expected arrival time of a photon, which contributes to the sifted key rate. This rate is

$$R_{\text{DC}} = R_{\text{rep}} \frac{1}{2} p_{\text{DC}} n_{\text{PD}},$$

with p_{DC} the probability of a dark count and n_{PD} the number of detectors. Here, the factor 1/2 is also due to the sifting. Together, the total sifted key rate is

$$R = R_{\text{AB}} + R_{\text{DC}}.$$

Now, let us calculate the error rate e, that is, the fraction of the rate R for which $X \neq Y$. First, the beam splitters may be slightly unbalanced or other optical imperfections may cause a photon to hit the wrong detector. This gives the optical error rate R_{opt}, which can be calculated as

$$R_{\text{opt}} = R_{\text{AB}} p_{\text{opt}},$$

10.3 Eavesdropping and secret key rate

with p_opt the probability that an incoming photon hits the wrong detector. Out of the total sifted key, this corresponds to an error probability of

$$e_\text{opt} = \frac{R_\text{opt}}{R} \approx p_\text{opt}.$$

Then, there are errors due to the dark counts, namely,

$$e_\text{DC} = \frac{1}{2}\frac{R_\text{DC}}{R} = \frac{1}{2}\frac{p_\text{DC} n_\text{PD}}{p_\text{TW}(1 - e^{-\mu})T\eta_\text{PD} + p_\text{DC} n_\text{PD}}.$$

The factor $1/2$ is due to the fact that about half of the time a dark count can occur at the correct detector, hence inducing no error. Notice that e_DC increases as T decreases, that is, when the fiber length increases.

Finally, the total error rate is $e = e_\text{opt} + e_\text{DC}$.

Note that, for sources based on parametric downconversion, multi-photon pulses also contribute to the error rate since they do not carry the same information. This is specific to such systems and will not be detailed further here.

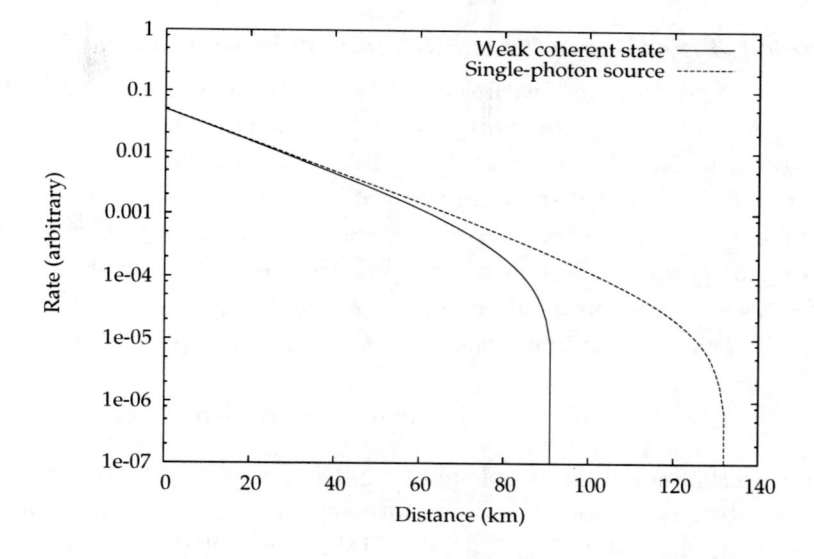

Fig. 10.7. Secret key rate as a function of the fiber length. We assume that the fiber losses are $0.25\,\text{dB}/\text{km}$ (at $1550\,\text{nm}$), $\eta_\text{PD} = 10\%$, $p_\text{DC} = 10^{-5}$, $p_\text{opt} = 0$, $p_\text{TW} = 1$ and $n_\text{PD} = 2$. For the single-photon source, we set $R_\text{rep} = 1$ in arbitrary units (e.g., Mbps). For weak coherent states, we assume $R_\text{rep} = 10.51$ and $\mu = 0.1$ in order to get a unit rate of non-empty pulses (i.e., $R_\text{rep}(1 - e^{-\mu}) = 1$) so as to allow a fair comparison with the single-photon source.

In Fig. 10.7, we plot the secret key rate as a function of the fiber length using typical values for the losses in the fiber and other parameters. Using these values, the range of BB84 using weak coherent states is about 90 km, while it raises to about 130 km using a single-photon source. Of course, the chosen parameters for this graph are arbitrary and the actual range may vary depending on the setup parameters.

The dark counts are the main reason for the limitation in range of the BB84 implementation. As the distance increases, the number of photons arriving safely at Bob's station decreases while the dark count rate, creating errors, increases. With respect to dark counts, weak coherent states and single photons are not equal, however. With weak coherent states, Alice and Bob do not know whether an empty or a non-empty pulse is sent. A dark count may occur during a time slot for which the pulse contains no photon. This is indistinguishable from a non-empty pulse arriving in one of Bob's detectors. For a single-photon source, Alice and Bob do know when a photon was sent, hence a smaller number of time slots must be considered. The number of dark counts is thus relatively lower.

10.3.4 Photon-number splitting attacks and decoy states

When Alice and Bob use weak coherent states instead of single-photon pulses, some pulses contain more than one photon. The use of weak coherent states is prone to PNS attacks: Eve can, in principle, measure the photon number without disturbing the state and decide what to do with the intercepted pulse. Note that, given the current technology, this would be quite difficult to do, but quantum mechanics does not prevent this, so let us nevertheless investigate this case. As a general assumption throughout this section, Eve's action may depend on the number of photons in the pulse sent by Alice.

Let us first describe the most systematic active PNS attack.

- For pulses that contain multiple photons, Eve keeps one of the photons and sends the others to Bob. Unlike an intercept-and-resend attack, this does not perturb the qubit transmitted to Bob since all the photons contain the same information in polarization or phase encoding. So when Alice reveals the encoding basis, Eve can make the measurement and obtain the secret key bit without introducing any errors.
- For pulses that contain only one photon, Eve discards them since she does not want to induce errors on Bob's side. The only effect of Eve's action is a decrease in the transmission probability T. So Eve's intervention induces losses, but she can compensate for them by sitting close to Alice's

10.3 Eavesdropping and secret key rate

station and by installing a virtually lossless fiber. So, we should assume that any losses can be due to Eve's attack.

Let us see what happens when Eve uses this strategy systematically. If the coherent states sent by Alice have μ photons in average, let us denote by P_1 and P_{2+} the ratio of non-empty pulses that contain only one and more than one photon(s), respectively, with

$$P_1 = \Pr[n=1|n>0] = \frac{\mu e^{-\mu}}{1-e^{-\mu}}, \text{ and}$$

$$P_{2+} = \Pr[n>1|n>0] = \frac{1-e^{-\mu}-\mu e^{-\mu}}{1-e^{-\mu}} = 1 - P_1.$$

Since with this attack all the one-photon pulses are discarded and all the multi-photon pulses are transmitted losslessly, the transmittance of this PNS attack is

$$T_{\min \text{ PNS}} = P_{2+}.$$

If no precautions are taken, the existence of this PNS attack may put a limit on the distance that one can achieve for secure QKD. If Alice and Bob use an optical fiber which is long enough so that $T \leq T_{\min \text{ PNS}}$, they have no way of distinguishing the expected losses in their optical fiber from a PNS attack. For instance, when $\mu = 0.1$, $T_{\min \text{ PNS}} \approx 0.049 \approx 13\,\text{dB}$ and this corresponds to a 52 km-long optical fiber, assuming 0.25 dB / km losses at 1550 nm.

Since Eve could use another eavesdropping strategy, let us further explore the PNS attacks in general and describe a way to quantify the amount of information that Eve is able to obtain.

The source at Alice's station produces coherent states $|\alpha e^{i\phi}\rangle$ whose absolute phase ϕ is randomized. Hence, it is characterized by the density matrix

$$\rho_\mu = \int \frac{d\phi}{2\pi} |\alpha e^{i\phi}\rangle\langle \alpha e^{i\phi}| \text{ with } \alpha = \sqrt{4N_0\mu}.$$

However, it can be shown formally that this density matrix can be diagonally rewritten in the Fock basis using photon number states $|n\rangle\langle n|$, namely,

$$\rho_\mu = \sum_n e^{-\mu} \frac{\mu^n}{n!} |n\rangle\langle n|$$
$$= e^{-\mu}|0\rangle\langle 0| + \mu e^{-\mu}|1\rangle\langle 1| + a\rho_{2+,\mu}.$$

Here, $\rho_{2+,\mu}$ is a density matrix that contains only multi-photon pulses and $a = 1 - e^{-\mu} - \mu e^{-\mu}$.

Alice thus sometimes sends no photon, sometimes one photon and sometimes a multi-photon pulse, following a Poisson distribution. However, she does not know the number of photons she sent. Eve, on her side, can make her attack dependent on the photon number, hence possibly causing different losses for the different number states. The transmittance of the source ρ_μ can be split as

$$T_\mu = \mu e^{-\mu} T_1 + a T_{2+,\mu}, \qquad (10.3)$$

with T_1 and $T_{2+,\mu}$ the transmittance of single-photon and multi-photon pulses, respectively. Since Alice does not perform a measurement of the photon number for the outgoing pulses, the legitimate parties are able to estimate T_μ but not the values of T_1 and $T_{2+,\mu}$.

The value of $T_{2+,\mu}$ would be interesting for Alice and Bob to know, since they would then be able to determine the ratio of tagged bits $\Delta = aT_{2+,\mu}/T_\mu \geq 0$, that is, the fraction of the key elements that are possibly known to the eavesdropper due to a PNS attack. This would allow the legitimate parties to take this information leakage into account during privacy amplification. Note that $\Delta = 1$ in the case of the systematic PNS attack where Eve blocks all single-photon states.

There is a powerful way for Alice and Bob to determine whether a PNS attack is going on and to obtain an upper bound on Δ; this is the *decoy state method* recently invented by Hwang [88]. The idea of this method is that Alice still sends ρ_μ for key bits, but she sometimes sends coherent states with a different average number of photons $\rho_{\mu'}$, called decoy states. Because decoy states are spread randomly and uniformly within all the pulses, Eve cannot know in advance which pulses are decoy states and which are regular states. After the transmission, Alice announces to Bob the position of decoy states so that they can determine $T_{\mu'}$, the transmittance for the decoy states. Decoy states are never used to transmit key elements, only to help determine Δ.

Let us consider the case where decoy states are more intense than regular states, that is, $\mu' \geq \mu$. The decoy state source $\rho_{\mu'}$ can be expressed as

$$\rho_{\mu'} = e^{-\mu'}|0\rangle\langle 0| + \mu' e^{-\mu'}|1\rangle\langle 1| + a\frac{\mu'^2 e^{-\mu'}}{\mu^2 e^{-\mu}}\rho_{2+,\mu} + b\rho_{3+,\mu'-\mu},$$

where $\rho_{3+,\mu'-\mu}$ is a density matrix that contains multi-photon pulses so as to complement those in $\rho_{2+,\mu}$ to obtain the Poisson distribution with average μ', and $b = 1 - e^{-\mu'} - \mu' e^{-\mu'} - a(\mu'^2 e^{-\mu'})/(\mu^2 e^{-\mu})$. The transmittance $T_{\mu'}$

of decoy states reads

$$T_{\mu'} = \mu' e^{-\mu'} T_1 + a \frac{\mu'^2 e^{-\mu'}}{\mu^2 e^{-\mu}} T_{2+,\mu} + b T_{3+,\mu'-\mu}. \qquad (10.4)$$

Eve's attack can only depend on the photon number but not on whether a pulse is a decoy state or not. This is why we can write the expression in Eq. (10.4) with the same coefficients T_1 and $T_{2+,\mu}$ as in Eq. (10.3).

Now, with the knowledge of T_μ and $T_{\mu'}$, Alice and Bob can determine an upper bound on Δ. Since $T_1 \geq 0$ and $T_{3+,\mu'-\mu} \geq 0$, we have:

$$T_{2+,\mu} \leq a^{-1} \frac{\mu^2 e^{-\mu}}{\mu'^2 e^{-\mu'}} T_{\mu'} \text{ or } \Delta \leq \frac{\mu^2 e^{-\mu}}{\mu'^2 e^{-\mu'}} \frac{T_{\mu'}}{T_\mu}. \qquad (10.5)$$

Note that, in practice, the upper bound in Eq. (10.5) above may be a bit pessimistic as only the inequalities $T_1 \geq 0$ and $T_{3+,\mu'-\mu} \geq 0$ are used. Better estimates of Δ can be obtained by using decoy states with other well-chosen intensities μ'', μ''', etc., not excluding intensities lower than μ – see for instance [110, 181]. Yet, the simple method with two intensities is enough to capture the philosophy of decoy states.

Most of the time, Eve is not present and no active PNS attack occurs. It is important that the decoy state method does not decrease the performance too much. This is fortunately the case. Note that even if no systematic PNS attack occurs, the use of weak coherent states implies that multi-photon pulses are present and may be eavesdropped. If the losses are only due to the optical fiber, the tagged bits ratio Δ is equal to the multi-photon rate $\Delta = P_{2+} \approx \mu/2$. For $\mu = 0.1$, $\Delta \approx 5\%$ and roughly speaking an extra 5% of information must be attributed to Eve, hence reducing the secret key rate after privacy amplification. However, a side benefit of decoy states is that it is also possible to use more intense pulses, say with $\mu = 0.5$, and thus to decrease the rate of empty pulses. Of course, more information is leaked via multi-photon pulses, but this is fully taken into account during privacy amplification. The increase in the rate of non-empty pulses outweighs the extra information gained by Eve. For rigorous derivations of the secret key rate in the presence of weak coherent states, please refer to Section 12.2.5.

10.4 Conclusion

In this chapter, I described the BB84 protocol and its physical implementation. The produced key elements can be used by the legitimate parties to distill a secret key. We also investigated the possible individual eavesdropping strategies against this protocol. Note that more general eavesdropping strategies are investigated in Chapter 12.

The key elements produced by BB84 are binary. Another family of protocols, which yield continuous key elements, is presented in the next chapter.

11
Protocols with continuous variables

The discrete modulation of quantum states as in BB84 stems fairly naturally from the need to produce zeroes and ones for the final secret key. But since secret-key distillation can also process continuous key elements, there is no reason not to investigate the continuous modulation of quantum states. In fact, continuous-variable protocols are fairly elegant alternatives to their discrete counterparts and allow for high secret key rates.

In this chapter, I describe two important QKD protocols involving continuous variables: first a protocol involving the Gaussian modulation of squeezed states and then a protocol with coherent states. These QKD protocols must be seen as a source of Gaussian key elements, from which we can extract a secret key. Finally, I detail the implementation of the protocol with coherent states.

11.1 From discrete to continuous variables

In the scope of continuous-variable QKD, we focus on quantum states that are represented in a continuous Hilbert space, of which their most notable examples are the coherent and squeezed states.

The BB84 protocol was designed with single photon states in mind. The coherent states, much easier to produce, are used only as an approximation of single photon states. Can coherent states serve as the base of a QKD protocol? Apart from the fact that coherent states contain in general more than one photon, they also form a family of states that are not orthogonal, in contrast to the states $\{|0\rangle, |1\rangle\}$ and $\{|+\rangle, |-\rangle\}$.

In 1992, Bennett [14] showed that the use of two non-orthogonal quantum states is a sufficient condition for QKD to be possible. This result opened the way for continuous-variable QKD. Although squeezed and coherent states may very well contain more than one photon, their non-orthogonality is sufficient for QKD to be possible.

Many of the first proposals for QKD with continuous variables involved entangled beams of light, showing EPR correlations of quadratures [9, 134, 144, 146, 164]. These states are difficult to produce, however. Some other protocols use more traditional prepare-and-measure procedures, where Alice prepares a state and sends it to Bob without keeping an entangled subsystem [40, 68, 82].

In [82], and to some extent in [68], the encoded key elements have a discrete nature. Cerf and coworkers proposed in [40] to use continuous – in fact, Gaussian – key elements, hence implying the continuous modulation of squeezed states. This protocol looks like a continuous-variable equivalent of BB84, yielding Gaussian key elements instead of binary ones.

Yet, the production of squeezed states is rather difficult. The use of coherent states, much simpler to produce, would be more interesting. Even though coherent states are particular cases of squeezed states, decreasing the squeezing $s \to 1$ in [40] makes the secret key rate go to zero.

The solution was found by Grosshans and Grangier, who designed the first protocol where coherent states are modulated in both quadratures simultaneously [74]; we call this protocol GG02. It also uses the idea of the Gaussian modulation: Alice generates coherent states of a light mode with Gaussian-distributed quadratures, and Bob's measurements are homodyne measurements. This protocol allows for facilitated implementations and high secret-key generation rates [77]; this follows from the fact that homodyne detection can operate faster than the photon detectors used for BB84. GG02 is detailed in Section 11.3. Note that many other protocols using coherent states have since then been discovered, e.g., [112, 165].

11.2 A protocol with squeezed states

Alice generates Gaussian key elements randomly and independently, denoted by the random variable X_A. In this protocol, there are two encoding rules, numbered by $i \in \{1, 2\}$. Alice randomly and independently chooses which encoding rule she uses for each key element. In general, the encoding rules may require different variances, and we thus denote by

$$X_{A,i} \sim N(0, \Sigma_{A,i}\sqrt{N_0})$$

the key elements obtained by appropriately scaling X_A before their encoding by rule i.

The idea of this protocol is as follows. The Heisenberg uncertainty princi-

11.2 A protocol with squeezed states

ple implies that it is impossible to measure with full accuracy both quadratures of a single mode, **x** and **p**. We exploit this property by encoding the key elements $X_{A,i}$ as a state squeezed either in **x** (case 1) or in **p** (case 2), in such a way that an eavesdropper, not knowing which of these two encodings is used, cannot acquire information without disturbing the state.

As a comparison, BB84 exploits the Heisenberg uncertainty principle with the two observables **Z** and **X**, as $[\mathbf{Z}, \mathbf{X}] \neq 0$. The states $\{|0\rangle, |1\rangle\}$ and $\{|+\rangle, |-\rangle\}$ are eigenstates of **Z** and **X**, respectively. For the current protocol, the eigenstates of **x** and **p** would be infinitely squeezed. Yet the squeezed states we use here are approximately the eigenstates of **x** and **p**. In this perspective, it is fair to say that this protocol acts as a continuous equivalent to BB84.

Let us now detail the two encoding rules.

- In case 1, Alice prepares a squeezed vacuum state such that the fluctuations of **x** are squeezed with parameter $s_1 < 1$, and applies a displacement of **x** by an amount equal to $X_{A,1}$, i.e., such that $\langle \mathbf{x} \rangle = X_{A,1}$, as depicted on Fig. 11.1. Hence Alice's encoding rule is $X_{A,1} \rightarrow |X_{A,1}, s_1\rangle$.
- Conversely, in case 2, Alice sends a squeezed state in **p** (i.e., with squeezing parameter $s_2 > 1$), whose displacement in **p** encodes the key element: $X_{A,2} \rightarrow |iX_{A,2}, s_2\rangle$.

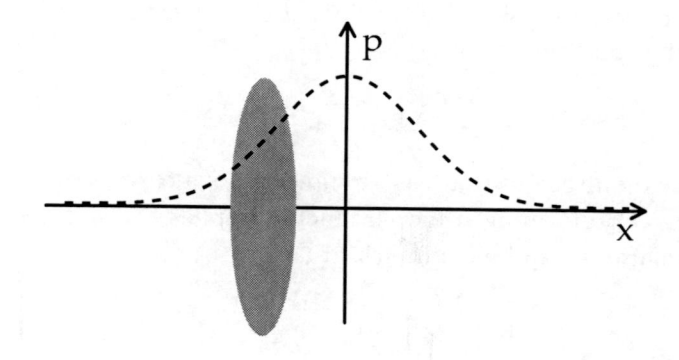

Fig. 11.1. Schematic description of the case 1 encoding rule. The squeezed states are modulated along the x axis. Their centers follow a Gaussian distribution, illustrated by the bell-shaped curve. The case 2 encoding rule is obtained by interchanging the x and p axes.

On his side, Bob measures either **x** or **p**, yielding the result $Y_{B,\mathbf{x}}$ or $Y_{B,\mathbf{p}}$, choosing at random which quadrature he measures. After sending a predefined number of squeezed states, Alice reveals to Bob the encoding rule for each squeezed state. They keep only the useful transmissions over the quantum channel, that is, they discard the key elements for which Alice used case 1 (or case 2) and Bob measured **p** (or **x**). For the remaining (sifted) key elements, we denote Bob's sifted measurements by Y_B.

From an observer's point of view, the mixed state that Alice sends in case 1 reads

$$\rho_1 = \int dx\, p_{N(0,\Sigma_{A,1}\sqrt{N_0})}(x)|x,s_1\rangle\langle x,s_1|.$$

In case 2, it becomes

$$\rho_2 = \int dp\, p_{N(0,\Sigma_{A,2}\sqrt{N_0})}(p)|ip,s_2\rangle\langle ip,s_2|.$$

We require the distribution of **x** measurement outcomes to be indistinguishable when case 1 or 2 is used by Alice, that is, $\rho_1 = \rho_2$. If this condition is fulfilled, Eve cannot obtain any indication on whether she is measuring a case 1 or case 2 squeezed state, whatever the statistics she accumulates. Let us call $\sigma_1^2 N_0 = \langle \mathbf{x}^2 \rangle = s_1 N_0 < N_0$ the intrinsic variance of the squeezed states in case 1 and $\sigma_2^2 N_0 = \langle \mathbf{p}^2 \rangle = N_0/s_2 < N_0$ in case 2. If case 1 is used, the outcomes of **x** measurements are distributed as a Gaussian of variance $(\Sigma_{A,1}^2 + \sigma_1^2)N_0$, since each squeezed state gives an extra contribution of σ_1^2 to the variance of the key elements $X_{A,1}$. If, on the contrary, a case 2 squeezed state is measured, then the outcomes of **x** measurements exhibit a Gaussian distribution of variance N_0/σ_2^2 as a result of the uncertainty principle. Thus, we impose the condition

$$\Sigma_{A,1}^2 + \sigma_1^2 = \frac{1}{\sigma_2^2}.$$

Similarly, the requirement that case 1 and case 2 squeezed states are indistinguishable when performing **p** measurements implies that $\Sigma_{A,2}^2 + \sigma_2^2 = 1/\sigma_1^2$. These two relations can be summarized as

$$1 + \frac{\Sigma_{A,1}^2}{\sigma_1^2} = 1 + \frac{\Sigma_{A,2}^2}{\sigma_2^2} = \frac{s_2}{s_1}. \tag{11.1}$$

Equation (11.1) also implies that the squeeze parameters s_1 and s_2 completely characterize the protocol.

11.2.1 Properties

Let us first analyze the case where there is no eavesdropping and the transmission is perfect. The transmission of the sifted key element X_A can be seen as a Gaussian channel, even in the absence of eavesdropping. Bob's measurement gives the random variable $Y_B = X_A + \epsilon$, where ϵ is the intrinsic noise of the squeezed state and has a Gaussian distribution. Referring to Eq. (3.1), the mutual information between X_A and Y_B is

$$I_{AB,0} \triangleq I(X_A, Y_B) = 2^{-1} \log\left(1 + \frac{\Sigma_{A,1}^2}{\sigma_1^2}\right) = 2^{-1} \log\left(1 + \frac{\Sigma_{A,2}^2}{\sigma_2^2}\right).$$

This mutual information $I_{AB,0}$ measures the number of bits that can be transmitted asymptotically per use of the channel with an arbitrary high fidelity for a given snr. This transmission rate can be shown to be attainable if the signal is Gaussian distributed, which is the case here. Note that, in the absence of squeezing, $s_1 = s_2 = 1$, $I_{AB,0} = 0$ and the protocol cannot work.

Let us now estimate the average photon number $\langle \mathbf{n} \rangle$ contained in each encoded state, assuming, for simplicity, that $\sigma_1 = \sigma_2 = \sigma$ and $\Sigma_{A,1} = \Sigma_{A,2} = \Sigma_A$, so that the same squeezing is applied on both quadratures. Without detailing the calculation (see [40]), one obtains

$$\langle \mathbf{n} \rangle = \frac{\sqrt{1 + \Sigma_A^2/\sigma^2} - 1}{2}.$$

Equivalently, the information $I_{AB,0}$ reads $I_{AB,0} = \log(2\langle \mathbf{n} \rangle + 1)$ bits, implying that the photon number increases exponentially with $I_{AB,0}$. We thus conclude that this QKD protocol clearly steps away from the single-photon and weak coherent state schemes.

11.2.2 Eavesdropping and secret key rate

For BB84, the optimal individual eavesdropping strategy implies the use of a phase-covariant cloning machine, as described in Section 10.3.2. For this squeezed state protocol, let us follow the same idea and describe an individual eavesdropping strategy with cloning machines such as those defined in [39]. Eve sits on the line between Alice and Bob, clones the state being transmitted, sends one of the clones to Bob and keeps the other one in a quantum memory. Since perfect cloning is impossible, Eve cannot acquire information without disturbing the transmitted quantum state. Let us analyze the amount of information Eve can acquire on the key as a function of the disturbance she induces.

As we shall see, the tradeoff between the information acquired by Bob and Eve in this protocol can be analyzed exactly. For that, we consider the strategy of Eve using the optimal (Gaussian) cloning machine for continuous quantum variables. More precisely, Eve makes two imperfect clones of the state sent by Alice, then sends a clone to Bob while she keeps the other one. Once Alice reveals the encoding rule, Eve measures her clone in the appropriate quadrature – this implies that Eve must keep her clones in a quantum memory until Alice reveals the encoding rules.

To analyze the information-theoretic balance between Bob and Eve, we use a class of asymmetric Gaussian cloning machines, which produce a different amount of noise on both quadratures for Bob and Eve. It is proven in [39] that the no-cloning inequality

$$\sigma_{B,\mathbf{x}}\sqrt{N_0}\sigma_{E,\mathbf{p}}\sqrt{N_0} \geq N_0$$

must hold and is saturated for this class of cloners, where $\sigma_{B,\mathbf{x}}\sqrt{N_0}$ and $\sigma_{E,\mathbf{p}}\sqrt{N_0}$ are the standard deviations of the errors that affect Bob's $\mathbf{x}$ measurements and Eve's $\mathbf{p}$ measurements, respectively. For example, in case 1, the outcomes of $\mathbf{x}$ measurements by Bob are distributed as a Gaussian of variance $(\sigma_1^2 + \sigma_{B,\mathbf{x}}^2)N_0$, since cloning-induced errors are added to the intrinsic fluctuations of the squeezed states. Similarly, a dual no-cloning uncertainty relation holds, connecting Bob's errors on $\mathbf{p}$ and Eve's errors on $\mathbf{x}$:

$$\sigma_{B,\mathbf{p}}\sqrt{N_0}\sigma_{E,\mathbf{x}}\sqrt{N_0} \geq N_0.$$

To calculate the information acquired by Bob and Eve, we characterize the cloners by two parameters χ and λ. We rewrite the cloning-induced error variances on Bob's side as $\sigma_{B,\mathbf{x}}^2 = \chi\lambda(\sigma_1^2/\alpha)$, where $\alpha = \sqrt{s_1/s_2}$, and $\sigma_{B,\mathbf{p}}^2 = \chi\lambda^{-1}(\sigma_2^2/\alpha)$, while the error variances on Eve's side are written as

$$\sigma_{E,\mathbf{x}}^2 = \chi^{-1}\lambda(\sigma_1^2/\alpha) \text{ and } \sigma_{E,\mathbf{p}}^2 = \chi^{-1}\lambda^{-1}(\sigma_2^2/\alpha).$$

Thus, χ characterizes the balance between Bob's and Eve's errors, and λ describes the balance between quadratures $\mathbf{x}$ and $\mathbf{p}$. Eve's measurement of the quadrature $\mathbf{p}$ in case 1 or $\mathbf{x}$ in case 2 is called Z.

Let us now express the mutual information between Alice and Bob and between Alice and Eve in quadratures $\mathbf{x}$ and $\mathbf{p}$. In case 1, the variance of Bob's measurements $Y_{B,\mathbf{x}}$ is $(\sigma_1^2 + \sigma_{B,\mathbf{x}}^2)N_0 = (1 + \chi\lambda/\alpha)\sigma_1^2 N_0$, while the distribution of $X_{A,1}$ has a variance $\Sigma_{A,1}^2 N_0$. Using Eq. (3.1) and (11.1), we obtain for Bob's information:

$$I_{AB,1} \triangleq I(X_{A,1}, Y_{B,\mathbf{x}}) = \frac{1}{2}\log\frac{1+\alpha\chi\lambda}{\alpha^2+\alpha\chi\lambda}. \qquad (11.2)$$

Similarly, using the variance of Z, $(\sigma_2^2 + \sigma_{E,\mathbf{p}}^2)N_0 = [1 + 1/(\alpha\chi\lambda)]\sigma_2^2 N_0$, we obtain for Eve's information:

$$I_{AE,1} \triangleq I(X_{A,1}, Z) = \frac{1}{2} \log \frac{1 + \alpha/(\chi\lambda)}{\alpha^2 + \alpha/(\chi\lambda)}. \tag{11.3}$$

Then, the balance between Bob's and Eve's information can be expressed by calculating the sum of Eq. (11.2) and (11.3):

$$I_{AB,1} + I_{AE,1} = \frac{1}{2} \log \alpha^{-2} = I_{AB,0}. \tag{11.4}$$

The information $I_{AE,1}$ acquired by Eve on the quadrature **p** is equal to the loss of mutual information on the quadrature **x** of the legitimate parties. Of course, the counterpart of Eq. (11.4) also holds when interchanging the quadratures (case 2), that is, $I_{AB,2} + I_{AE,2} = I_{AB,0}$.

For simplicity, we now consider a symmetric case, where $I_{AB,1} = I_{AB,2} \triangleq I_{AB}$ and $I_{AE,1} = I_{AE,2} \triangleq I_{AE}$. The information loss $I_{AB,0} - I_{AB}$ can be viewed as a disturbance measure as it turns out to be an upper bound on the information that might be gained by a potential eavesdropper,

$$I_{AE} \leq I_{AB,0} - I_{AB}.$$

The above description and analysis of this protocol characterizes the random variables X_A, Y_B and Z obtained by Alice, Bob and Eve, respectively. In particular, it gives an upper bound on $I_{AE} = I(X_A; Z)$. If we consider that the reconciled key Ψ is determined by X_A, the secret key rate is lower bounded by $\Delta I \triangleq I_{AB} - I_{AE}$, as detailed in Chapter 6. Consequently, the amount of secret key bits that can be generated by this method is bounded from below by $\Delta I \geq 2I_{AB} - I_{AB,0}$. This protocol is guaranteed to generate a non-zero secret key rate provided that the quality of the quantum channel is such that $I_{AB} > I_{AB,0}/2$, that is, if the losses are less than 3 dB.

Unfortunately, squeezed states are more difficult to produce than coherent states and the above protocol does not work in the particular case where $s_1 = s_2 = 1$. We thus now present a protocol based on coherent states.

11.3 A protocol with coherent states: the GG02 protocol

In the GG02 protocol, Alice uses only one encoding rule, but she encodes two different key elements, one of which will be discarded by Bob. Alice randomly and independently generates Gaussian key elements, denoted by the random variables $X_{A,1}$ and $X_{A,2}$.

The encoding of $X_{A,1} \sim N(0, \Sigma_A \sqrt{N_0})$ and $X_{A,2} \sim N(0, \Sigma_A \sqrt{N_0})$ consists in creating a coherent state whose displacement in **x** encodes $X_{A,1}$ and whose displacement in **p** encodes $X_{A,2}$, as illustrated in Fig. 11.2. More precisely, the encoding rule is:

$$(X_{A,1}, X_{A,2}) \to |X_{A,1} + iX_{A,2}\rangle. \tag{11.5}$$

The idea of this protocol is, again, that the Heisenberg uncertainty principle prevents one from measuring with full accuracy both quadratures of a single mode, **x** and **p**. On his side, Bob randomly chooses to measure either **x** or **p**. The result of his measurement is denoted by Y_B. When he measures **x**, the value Y_B is (a-priori) correlated to $X_{A,1}$, so Alice and Bob discard $X_{A,2}$ and set $X_A = X_{A,1}$. Conversely, when Bob measures **p**, Alice and Bob discard $X_{A,1}$ and set $X_A = X_{A,2}$.

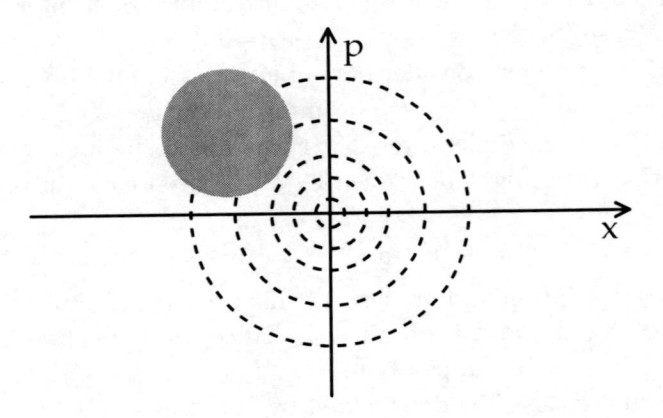

Fig. 11.2. Schematic description of the encoding rule. The coherent states, such as the one illustrated in the upper left quadrant, are modulated along both axes. Their centers follow a bivariate Gaussian distribution, illustrated by the concentric circles.

The protocol BB84 and the protocol of Section 11.2 both rely on the sifting of uncorrelated measurements. This protocol is different in the sense that no quantum state is discarded, but instead two pieces of information are encoded, one of which is discarded.

From an observer's point of view, the mixed state that Alice sends is

$$\rho = \int dx \int dp \, p_{N(0,\Sigma_A \sqrt{N_0}) \times N(0,\Sigma_A \sqrt{N_0})}(x,p) |x + ip\rangle\langle x + ip|. \tag{11.6}$$

11.3.1 Properties

Let us first analyze the case of a perfect quantum channel. Like the protocol of Section 11.2, the transmission of X_A to Y_B behaves like a Gaussian channel. The variance of the signal measured by Bob comprises the contribution of Alice's modulation $\Sigma_A^2 N_0$ and of the intrinsic fluctuations of the coherent state N_0, hence the variance of Y_B is $(\Sigma_A^2 + 1)N_0 = \Sigma^2 N_0$, with $\Sigma^2 = \Sigma_A^2 + 1$.

Let us now consider the case of a channel with losses, that is, an attenuation channel. This channel can be simply modeled as being composed of a beam splitter of attenuation $0 \leq G \leq 1$, through which Alice's coherent states pass. Let $\mathbf{x}_A$ be the quadrature amplitude of the state sent by Alice and let this state hit a beam splitter, on which a vacuum state $|0\rangle$ with amplitude quadrature $\mathbf{x}_0$ is input on the other side of the beam splitter. The output quadrature amplitude $\mathbf{x}_B$ transmitted to Bob is expressed as $\mathbf{x}_B = \sqrt{G}\mathbf{x}_A + \sqrt{1-G}\mathbf{x}_0 = \sqrt{G}(\mathbf{x}_A + \sqrt{\chi_B}\mathbf{x}_0)$ with $\chi_B = (1-G)/G$.

More generally, a quantum Gaussian channel can be obtained by inputting a Gaussian modulation of coherent states (instead of vacuum states) to the other side of the beam splitter. In this case, the output quadrature amplitude reads $\mathbf{x}_B = \sqrt{G}(\mathbf{x}_A + \sqrt{\chi_B}\mathbf{x}_0)$ with $\chi_B = (1-G)/G + \epsilon$. The added noise variance χ_B thus contains a contribution $(1-G)/G$ due to the attenuation of the channel, and a contribution $\epsilon \geq 0$ due to any other added noise.

In the case of a quantum Gaussian channel, the variance of the signal measured by Bob is scaled by G and comprises the intrinsic fluctuations N_0, the added noise contribution $\chi_B N_0$ and Alice's modulation $\Sigma_A^2 N_0$, hence the variance of Y_B is $\Sigma_B^2 N_0 = G(\Sigma_A^2 + 1 + \chi_B)N_0 = G(\Sigma^2 + \chi_B)N_0$. Consequently, the information shared by Alice and Bob is

$$I_{AB} \triangleq I(X_A; Y_B) = \frac{1}{2}\log\left(1 + \frac{\Sigma_A^2}{\chi_B + 1}\right).$$

11.3.2 Eavesdropping and secret key rate

It is fairly natural to say that the reconciled key Ψ is determined by X_A, the key elements of Alice's modulation. We call this *direct reconciliation* (DR) and the secret key rate is given by $\Delta I_d = I(X_A; Y_B) - I(X_A; Z)$. For direct reconciliation, we follow the convention that $X = X_A$, $Y = Y_B$, Claude is in fact Alice and Dominique is Bob. The original protocol proposed in [74] uses this expression to calculate the secret key rate and to determine the optimal individual eavesdropping strategy. As I will detail below, the disadvantage is that the secret key rate drops to zero when the channel attenuates the signal power by half or more ($G \leq 1/2$).

Of course, Theorem 3 of Section 6.4 says that one can also determine the reconciled key Ψ from Bob's measurements. This case yields a secret key rate of $\Delta I_r = I(X_A; Y_B) - I(Y_B; Z)$ and is called *reverse reconciliation* (RR). So, the convention is $X = Y_B$, $Y = X_A$, Claude is in fact Bob and

Dominique is Alice. The optimal individual eavesdropping strategy can also be determined in this case, with the property that $\Delta I_r > 0$ for any attenuation channel ($\chi_B = (1 - G)/G$, $\epsilon = 0$) whenever the attenuation $G > 0$, even when it is less than a half [75, 77].

We can now analyze the two different cases, first direct reconciliation and then reverse reconciliation.

Direct reconciliation

In the case of direct reconciliation, the optimal individual Gaussian eavesdropping strategy consists in intercepting (reflecting) a part of the signal with a beam splitter, while transmitting the rest of the signal to Bob. Eve keeps the reflected state in a quantum memory, until she knows which quadrature is measured by Bob; we can thus assume that Eve always measures the right quadrature. The result of her measurement is called Z.

With a similar reasoning to that in Section 11.2.2, Bob and Eve cannot both acquire all the information about the key elements; the variance of Bob on the quadrature $\mathbf{x}$ and of Eve on the quadrature $\mathbf{p}$ together verify a no-cloning uncertainty relation. If we denote by χ_E the noise of the channel going from Alice to Eve, it must verify $\chi_B \chi_E \geq 1$.

Even if Alice and Bob know they have a lossy line with attenuation G, a powerful eavesdropper could replace the lossy line by a perfect line and insert a beam splitter with transmittance G; Alice and Bob would not be able to tell the difference. We must thus assume that Eve is able to intercept a part of the signal, on which she sees the noise variance $\chi_E = 1/\chi_B$.

As calculated in [74, 76], the information shared by Alice and Eve is

$$I_{AE} \triangleq I(X_A; Z) = \frac{1}{2} \log \left(1 + \frac{\Sigma_A^2}{\chi_E + 1}\right) = \frac{1}{2} \log \left(1 + \frac{\Sigma_A^2}{\frac{1}{\chi_B} + 1}\right),$$

and the secret key rate is

$$\Delta I_d = I_{AB} - I_{AE} = \frac{1}{2} \log \left(\frac{\Sigma^2 + \chi_B}{\chi_B \Sigma^2 + 1}\right).$$

With an attenuation channel, ΔI_d drops to zero when $G \leq 1/2$, that is, when the losses are above 3 dB.

Reverse reconciliation

In the case of reverse reconciliation, the optimal individual Gaussian eavesdropping strategy consists in using an *entangling cloning machine* [75, 77]. As depicted in Fig. 11.3, an entangling cloning machine not only reflects a part of the signal and transmits the rest of it (as for direct reconciliation)

11.3 A protocol with coherent states: the GG02 protocol

but also generates an entangled state, inserting half of it into the signal to Bob and measuring the other half of it. Eve can thus try to influence Bob's signal. We can again assume that Eve uses quantum memory and thus measures the right quadrature. We denote by Z_m Eve's measurement of the signal coming from Alice and by Z_i the noise induced by Eve to Bob's signal. Hence, $Z = (Z_i, Z_m)$.

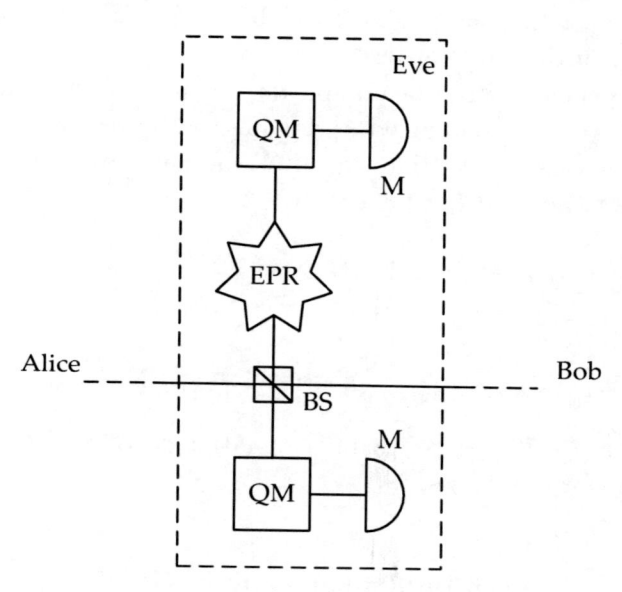

Fig. 11.3. Eve uses an entangling cloning machine for optimal eavesdropping of GG02 with RR. The entangling cloning machine comprises a beam splitter (BS) through which Eve injects half of an entangled pair (EPR, upper part of the figure). The other half is kept in a quantum memory (QM) for later measurement (M). The beam splitter also extracts a fraction of the coherent states sent by Alice (lower part). The extracted state is also kept in a quantum memory (QM) and measured (M) when Bob reveals which observable he measured.

In this context, an entangled state is such that the quadrature of one half is correlated to the quadrature of the other half. Let $\mathbf{x}_1, \mathbf{p}_1$ be the quadrature amplitudes of the first half (the one Eve keeps) and $\mathbf{x}_2, \mathbf{p}_2$ the quadrature amplitudes of the second half (the one Eve inputs into the beam splitter). Since the commutator $[\mathbf{x}_1-\mathbf{x}_2, \mathbf{p}_1+\mathbf{p}_2] = 0$, Eve can simultaneously correlate the quadratures $\mathbf{x}_2, \mathbf{p}_2$ of the state she sends to Bob to the quadratures she keeps in her quantum memory $\mathbf{x}_1, \mathbf{p}_1$. However, the commutator

$$[\mathbf{x}_1 - \mathbf{x}_2, \mathbf{p}_2] = 2iN_0$$

implies that the correlation between the quadrature $\mathbf{x}$ in the two halves cannot be perfect, unless with infinite energy. The same argument applies to the correlations in **p**. Because of the uncertainty relation $\langle(\mathbf{x}_1 - \mathbf{x}_2)^2\rangle\langle\mathbf{p}_2^2\rangle \geq N_0^2$, Eve has to trade better correlations $\langle(\mathbf{x}_1 - \mathbf{x}_2)^2\rangle \leq N_0$ with a larger variance $\langle\mathbf{p}_2^2\rangle \geq N_0$ and thus a larger excess noise as seen by the legitimate parties.

Note that inputting a vacuum state into the beam splitter is a special case of the ECM, with $\langle\mathbf{x}_2^2\rangle = \langle\mathbf{p}_2^2\rangle = N_0$. The beam splitter attack is thus a particular case of the ECM attack.

Eve has the choice of the beam splitter balance (induced losses) and of the correlations of the entangled state (induced excess noise). Whatever she chooses, one can show that the best she can do in terms of mutual information is to acquire I_{BE} bits, with

$$I_{\text{BE}} = \frac{1}{2}\log(G^2(\Sigma^2 + \chi_{\text{B}})(\Sigma^{-2} + \chi_{\text{B}})),$$

and

$$\Delta I_{\text{r}} = I_{\text{AB}} - I_{\text{BE}} = \frac{1}{2}\log(G^2(1 + \chi_{\text{B}})(\Sigma^{-2} + \chi_{\text{B}})),$$

as calculated in [75, 76, 77]. With an attenuation channel, $\Delta I_{\text{r}} > 0$ whenever $G > 0$, that is, for any losses.

11.4 Implementation of GG02

Figure 11.4 shows the general setting for the implementation of GG02. Alice produces intense coherent states, which enter a beam splitter. The beam splitter is not balanced and only a small fraction of the intensity is directed to a phase and amplitude modulator. The modulator allows the coherent state to have its center chosen arbitrarily in the complex plane, as required by the encoding rule in Eq. (11.5). The intense part of the beam is called the *local oscillator* (LO) and serves as a phase reference.

Both beams are transmitted to Bob. The local oscillator does not depend on any secret information, while the modulated coherent states carry the quantum modulated key elements.

On Bob's side, both beams enter the homodyne detector. They interact through a balanced beam splitter. The two output beams are directed into photon detectors, which measure the intensity of the beam. The difference of the measured intensities is proportional to a quadrature amplitude. Let me detail how it works.

The local oscillator consists in an intense coherent state $|\alpha_{\text{LO}}\rangle$ with amplitude $|\alpha_{\text{LO}}| \gg \sqrt{N_0}$. The phase θ of the local oscillator ($\alpha_{\text{LO}} = |\alpha_{\text{LO}}|e^{i\theta}$)

11.4 Implementation of GG02

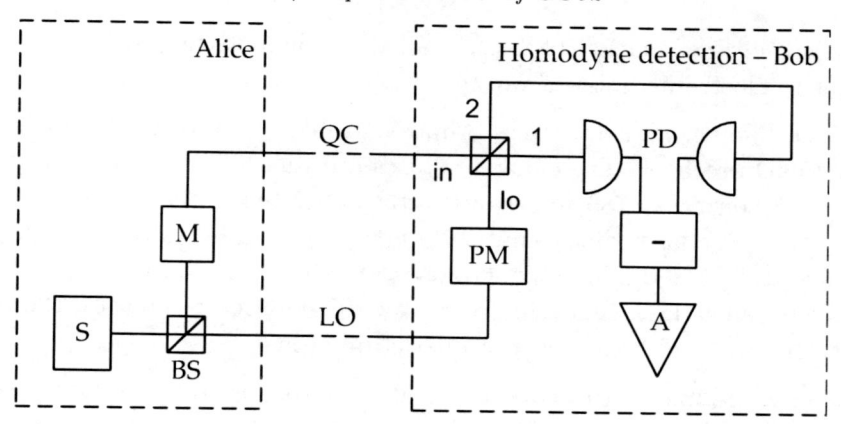

Fig. 11.4. Alice's station comprises a source of coherent states (S), often a laser, a beam splitter (BS) and a modulator (M). The beams are transmitted as the quantum channel (QC) and as the local oscillator (LO). In Bob's station, the local oscillator undergoes phase modulation (PM) and interacts with the other beam in a beam splitter. The output beam intensities are measured in photon detectors (PD). The resulting signal is subtracted (−) and amplified (A).

is determined by the phase modulator. The local oscillator enters the beam splitter at the lo arm, while the state ρ to measure enters in the in arm. At the output of the beam splitter, the number of photons in both arms are represented by the following operators:

$$\mathbf{n}_1 = \frac{1}{2}(\mathbf{a}_{lo}^\dagger + \mathbf{a}_{in}^\dagger)(\mathbf{a}_{lo} + \mathbf{a}_{in}),$$

$$\mathbf{n}_2 = \frac{1}{2}(\mathbf{a}_{lo}^\dagger - \mathbf{a}_{in}^\dagger)(\mathbf{a}_{lo} - \mathbf{a}_{in}).$$

The difference of intensity measured by the two detectors is

$$\Delta \mathbf{I} = 4N_0(\mathbf{n}_1 - \mathbf{n}_2) = 4N_0(\mathbf{a}_{lo}^\dagger \mathbf{a}_{in} + \mathbf{a}_{in}^\dagger \mathbf{a}_{lo}).$$

Since the local oscillator is very intense, it can be approximated by a classical state. Hence, we can rewrite the intensity as follows:

$$\Delta \mathbf{I} \approx 4N_0 \left(\frac{\alpha_{LO}^*}{2\sqrt{N_0}} \mathbf{a}_{in} + \frac{\alpha_{LO}}{2\sqrt{N_0}} \mathbf{a}_{in}^\dagger \right)$$
$$= |\alpha_{LO}| 2\sqrt{N_0} (e^{-i\theta} \mathbf{a}_{in} + e^{i\theta} \mathbf{a}_{in}^\dagger)$$
$$= 2|\alpha_{LO}| \mathbf{x}_\theta,$$

with $\mathbf{x}_\theta = \cos\theta \, \mathbf{x} + \sin\theta \, \mathbf{p}$.

Homodyne detection allows Bob to measure $\mathbf{x}$ when $\theta = 0$ or $\mathbf{p}$ when $\theta = \pi/2$. In practice, however, the measurement is not perfect. The signal undergoes an attenuation of G_{hom}, called the homodyne efficiency, and is subject to electronic noise of variance $\chi_{\text{elec}} N_0$.

In [77], Wenger and coworkers propose a table-top implementation of the whole QKD protocol. The coherent states are generated with a continuous-wave laser source at 780 nm wavelength associated with an acousto-optic modulator. The laser diode emits 120 ns (full-width at half-maximum) coherent states at 800 kHz. Light pulses are then split by a beam-splitter, one beam being the local oscillator (1.3×10^8 photons per pulse), the other being Alice's signal beam, with up to 250 photons.

The modulation of the coherent states in the $(x, p) = (r\cos\phi, r\sin\phi)$ plane is carried out in two steps. First, the amplitude r of each pulse is arbitrarily modulated by an integrated electro-optic modulator. However, owing to the unavailability of a fast phase modulator at 780 nm, the phase ϕ is not randomly modulated but scanned continuously from 0 to 2π. No genuine secret key can be distributed, strictly speaking, but random permutations of Bob's data are used to provide realistic data. All voltages for the amplitude and phase modulators are generated by an acquisition board connected to a computer. Although all discussions assume the modulation to be continuous, digitized voltages are used in practice. The modulation voltage is produced using a 16-bit converter, which was found to be enough for the modulation to be considered as continuous [76, 185].

Bob's station comprises a homodyne detection with an overall homodyne detection efficiency G_{hom} that ranges between 0.81 and 0.84. The electronic noise variance is $\chi_{\text{elec}} N_0 = 0.33 N_0$.

Instead of a table-top implementation, Lodewyck, Debuisschert, Tualle-Brouri and Grangier propose an implementation of GG02 in fiber optics [111]. The laser diode generates 100 ns coherent states at 1550 nm wavelength. The modulation on Alice's side is performed by computer-driven electro-optics amplitude and phase modulators and is truncated to four standard deviations.

The phase modulator of Bob's homodyne detection allows him to select the desired quadrature at 1 MHz. Hence, the repetition rate can be up to 1 MHz. Due to fiber connectors, the homodyne detection is currently $G_{\text{hom}} = 0.60$. The excess noise variance is about $0.06 N_0$.

In this implementation, the modulated coherent states and the local oscillator are sent over two separate fibers. To avoid relative polarization and phase drifts between the two fibers, experiments are under way to make both signals travel in the same fiber using time multiplexing.

11.4.1 Parameter estimation

After the quantum transmission, Alice and Bob have to reveal a random subset of their key elements to evaluate the transmission G_{line} and the total added noise variance $\chi_B N_0$. The variance of Bob's measurement has five contributions: the signal variance, the intrinsic fluctuations N_0, the channel noise $\chi_{\text{line}} N_0$, the electronic noise of Bob's detector $\chi_{\text{elec}} N_0$, and the noise due to imperfect homodyne detection efficiency $\chi_{\text{hom}} N_0 = (1 - G_{\text{hom}}) N_0 / G_{\text{hom}}$.

The total noise variance is

$$\chi_B N_0 = \chi_{\text{line}} N_0 + \frac{\chi_{\text{elec}} + \chi_{\text{hom}}}{G_{\text{line}}} N_0.$$

The two detection noises $\chi_{\text{elec}} N_0$ and $\chi_{\text{hom}} N_0$ originate from Bob's detection system, so one may reasonably assume that they do not contribute to Eve's knowledge; this is the *realistic approach*. In contrast, in the *paranoid approach*, one should assume that the detection noises are also controlled by Eve, which give her an additional advantage. Note that the realistic approach still assumes that Eve has unlimited computing resources and no technical limitations (perfect quantum memories, perfect fibers, perfect entanglement sources) – this is rather conservative for a realistic approach.

For each data burst, the values X_A coming from Alice and Y_B measured by Bob are processed in the following way. First, the variances of the signals, $\sigma^2[X_A]$ and $\sigma^2[Y_B]$, and the correlation coefficient $\rho = \rho[X_A, Y_B]$ are estimated. From this, we can deduce the modulation variance $\Sigma_A^2 = \sigma^2[X_A]/N_0$, the losses due to the transmission line or to an eavesdropper $G_{\text{line}} = \rho^2 \sigma^2[Y_B]/\sigma^2[X_A]$ and the total added noise variance, which amounts to $\chi_B = \sigma^2[X_A](\rho^{-2} - 1)/N_0 - 1$.

To encompass both the realistic and the paranoid approaches, the mutual information can be expressed as

$$I_{AE} = \frac{1}{2} \log \left(1 + \frac{\Sigma_A^2}{1 + \chi_{B,E}^{-1}} \right),$$

and

$$I_{BE} = \frac{1}{2} \log \frac{G_{\text{line}} G_{\text{extra}} (\Sigma^2 + \chi_B)}{\chi_{B,B} G_{\text{line}} + 1/G_{\text{line}} G_{\text{extra}} (\chi_{B,E} + \Sigma^{-2})},$$

with $G_{\text{extra}} = G_{\text{hom}}$, $\chi_{B,E} = \chi_B$ and $\chi_{B,B} = 0$ in the paranoid approach, or $G_{\text{extra}} = 1$, $\chi_{B,E} = \chi_{\text{line}}$ and $\chi_{B,B} = (\chi_{\text{elec}} + \chi_{\text{hom}})/G_{\text{line}}$ in the realistic approach [76, 185].

The results of [77] are presented in Table 11.1 for the paranoid approach and in Table 11.2 for the realistic approach. The paranoid assumptions are so stringent that the secret key rate drops to zero even for weak losses. In the realistic approach, the secret key rate (in reverse reconciliation) is non-zero for up to about 6 dB losses. The reverse reconciliation thus outperforms the direct reconciliation, which does not allow one to obtain a secret key when the losses are above 3 dB. The 800 kHz repetition rate yields the rates in kbps. These are theoretical secret key rates, which do not take into account the efficiency of reconciliation.

Table 11.1. *Results in the case of the paranoid approach, for both direct and reverse reconciliations.*

Σ_A^2	G_{line}	I_{AB}	I_{AE}	I_{BE}	ΔI_d	(kbps)	ΔI_r	(kbps)
40.7	1.00 (0.0 dB)	2.39	1.96	1.88	0.43	340	0.51	410

Table 11.2. *Results in the case of the realistic approach, for both direct and reverse reconciliations.*

Σ_A^2	G_{line}	I_{AB}	I_{AE}	I_{BE}	ΔI_d	(kbps)	ΔI_r	(kbps)
40.7	1.00 (0.0 dB)	2.39	0.00	0.00	2.39	1920	2.39	1920
37.6	0.79 (1.0 dB)	2.17	1.49	1.23	0.69	540	0.94	730
31.3	0.68 (1.7 dB)	1.93	1.69	1.30	0.24	190	0.63	510
26.0	0.49 (3.1 dB)	1.66	1.87	1.20	-		0.46	370
42.7	0.26 (5.9 dB)	1.48	2.53	1.38	-		0.10	85

11.5 GG02 and secret-key distillation

In [77], the experiment also includes an implementation of the secret-key distillation of the key elements obtained from the experiment detailed above. This includes the estimation of the parameters (see Section 11.4.1), the reconciliation (see Section 9.3.3) and the privacy amplification (see Chapter 7). It only remains to show how to connect all these parts and to display the results.

11.5.1 Privacy amplification

In this chapter, let us restrict ourselves to the case of individual attacks, i.e., Eve is only allowed to interact with a single coherent state at a time and she measures her probe as soon as Bob's chosen quadratures are revealed. Actually, a further requirement for the eavesdropping to be Gaussian [77] is not detailed here, as it turns out that this requirement is not necessary [79]. General eavesdropping strategies are discussed in Sections 12.1 and 12.3.

To encompass both direct and reverse reconciliations when talking about secret-key distillation, we use the Claude/Dominique convention, with X for Claude's variable and Y being Dominique's variable:

$$\text{Direct reconciliation:} X = X_A, Y = Y_B,$$
$$\text{Reverse reconciliation:} X = Y_B, Y = X_A.$$

From the assumption of individual eavesdropping, the value of $l \times I(X;Z)$, with l the block size, is a good estimate of the number of bits known to Eve, in the sense of privacy amplification using a universal family of hash functions as defined in Section 6.3.1. To apply such a technique, Claude and Dominique must estimate the eavesdropper's uncertainty in terms of order-2 Rényi entropy. When the block size goes to infinity, the Rényi and Shannon entropies become equal, conditionally on the typicality of the random variables; the probability of the random variables being typical can be made arbitrarily close to 1 [122]. Hence, for large block sizes, one must remove at least $I(X;Z)$ bits per key element during privacy amplification to make the final key secret.

Ideally, one should calculate the rate of convergence of the exact number of bits to remove by privacy amplification towards $l \times I(X;Z)$. Unfortunately, this turns out to be a difficult problem. The rate of convergence requires the knowledge of the exact eavesdropping strategy, i.e., the joint distribution of Claude's, Dominique's and Eve's symbols. However, there is no guarantee that Eve uses the optimal eavesdropping (in the sense that it maximizes $I(X;Z)$), and the number of bits to remove may be different for a finite block size.

In addition, the estimation of $I(X;Z)$ uses only a finite number of samples and should include a safety margin, so as to properly upper bound the information Eve could really acquire. We only calculate the expected number of secret key bits, hence neglecting the statistical fluctuations of $I(X;Z)$. The statistical fluctuations can be estimated with the bootstrap technique – see, e.g., [55].

11.5.2 Problem with interactive reconciliation

In Section 9.3.3, we showed the reconciliation efficiencies for the data generated by the experiment of [77]. The BCP used is the binary interactive error correction protocol Cascade. Rigorously, the size of the reconciliation messages $|M|$ must take into account the bits sent by both parties. However, this does not take advantage of the high efficiency of Cascade and yields poor reconciliation efficiencies and low secret key rates. I shall now further analyze the problem and then propose two solutions.

Let us assume that we are correcting slice i. As explained in Section 8.3.2, Cascade discloses the parities RS and RE of the vectors

$$S = S_i(X_{1...l}) \text{ and } E = E_i(Y_{1...l}, S_{1...i-1}(X_{1...l}))$$

of size l each $(S, E \in \mathrm{GF}(2)^l)$, where R is a binary matrix of size $r \times l$ for some integer r. Alice and Bob thus always communicate the parities calculated over identical subsets of bit positions.

If S and E are balanced and are connected by a binary symmetric channel, the parities RS give Eve r bits of information on S, but RE does not give any extra information since it is merely a noisy version of RS. Stated otherwise, $S \to RS \to RE$ is a Markov chain, hence only $r \approx l(1+\epsilon)h(e_i)$ bits are disclosed, which is not far away from the ideal $lh(e_i)$, where e_i is the bit error rate.

However, in the more general case where Eve gathered in Z some information on S and E by tapping the quantum channel, $S|Z \to RS|Z \to RE|Z$ does not necessarily form a Markov chain. Instead, we must upper bound Eve's information with the number of bits disclosed by both parties as if they were independent, $|M| = 2r \approx 2l(1+\epsilon)h(e_i)$, which is unacceptably high.

Such a penalty is an effect of interactivity, as both Claude and Dominique disclose some information. This can however be reduced by noticing that RS and RE can also be equivalently expressed by RS and $R(S+E)$. The first term RS gives information directly on the reconciled key Ψ via $S_C(X)$, where C is the set of slice indices for which Cascade was used. Intuitively, the second term $R(S+E)$ contains the discrepancies of Claude's and Dominique's bits, which is mostly noise, and does not contribute much to Eve's knowledge. Yet, the non-uniformity of the intervals in sliced error correction implies that errors may be more likely for some values than for others; there is thus some correlation between $R(S+E)$ and S. This must be evaluated explicitly.

There are, therefore, two solutions to this problem. First, we can try to explicitly evaluate Eve's information caused by the interactivity of Cascade. Second, we can replace Cascade by a one-way BCP. These two solutions are now investigated.

11.5.3 Evaluation of interactivity costs

The information gained by Eve after the quantum transmission and the reconciliation is $I(\Psi; Z, M)$, where Ψ is the reconciled key, Z is the information gained by Eve during the quantum transmission and M are the reconciliation messages.

The information $I(\Psi; Z, M)$ can be split as

$$I(\Psi; Z, M) = I(\Psi; Z) + I(\Psi; M|Z).$$

The variable M represents both the messages from Claude to Dominique and vice versa. The messages of the former kind depend on the slice bits and are of the form RS, while those of the latter kind depend on the estimators and are of the form RE.

The pair (RS, RE) can be equivalently written as $(RS, R(S + E))$. The protocol Cascade is designed in such a way that $R(S + E)$ contains enough parities to be able to correct all the errors. Furthermore, the decision to flip a bit depends only on the differences between Claude's and Dominique's parities. Hence, $R(S+E)$ contains all the information necessary to determine the error pattern $S + E$, and the pair (RS, RE) can be equivalently written as $(RS, S + E)$.

We can thus further split $I(\Psi; Z, M) = I(\Psi; Z) + I(\Psi; RS, S + E|Z)$ into

$$I(\Psi; Z, M) = I(\Psi; Z) + I(\Psi; S + E|Z) + I(\Psi, RS|S + E, Z)$$
$$\leq I(\Psi; Z) + I(\Psi; S + E|Z) + |M_{\text{ow}}|.$$

The first term is the information that Eve could gain on the quantum channel and can be upper bounded by $l \times I_{\text{AE}}$ or $l \times I_{\text{BE}}$, depending on whether we use DR or RR. The third term $|M_{\text{ow}}|$ is the number of bits disclosed one-way from Claude to Dominique. Finally, we need to evaluate the second term $I(\Psi; S+E|Z)$, which gives us the cost of interactivity, that is, the additional information disclosed by Dominique when using Cascade.

One solution is to use a part of the final key of the previous block to encrypt the reconciliation with the one-time pad. Both parties use the same key bits when encrypting the parity over a given subset of positions. Hence, an observer cannot determine the value of the parities but can tell when there is a difference. This solution is derived from [109] (see also Section 12.2.4) and is equivalent to the one I present now in terms of secret-key rate.

The evaluation of $I(\Psi; S + E|Z)$ requires the joint description of S, E and Z. For this to be possible, we need to assume that a particular kind of attack is being used. We thus assume that the entangling cloning machine

(ECM) is used; the beam splitter attack is a particular case and is thus automatically considered.

Care must now be taken, as the ECM is not necessarily the optimal individual eavesdropping strategy in this new setting. The ECM is optimal for Eve to maximize $I_{\text{AE or BE}} = I(X; Z)$. There may also be other eavesdropping strategies such that $I(X; Z) + I(\Psi; S + E|Z_{\text{ECM}})$ is maximized, where $I(\Psi; S + E|Z_{\text{ECM}})$ is evaluated with the assumption that the ECM is used even though it is not necessarily true. In this case, Eve gets a lower $I(X; Z)$ but gets more information overall than Alice and Bob expect. Maximizing $I(X; Z) + I(\Psi; S + E|Z)$ for any eavesdropping strategy seems to be very difficult. Hence, the method presented here is subject to the assumption that Eve actually uses an ECM.

To evaluate $I(\Psi; S+E|Z)$, one must integrate $p_Z(z)I(\Psi; S+E|Z = z)$ over all possible values of Z. In the case of the ECM, Eve knows $Z = (Z_i, Z_m)$, where Z_i is the noise induced at Bob's station via her EPR pair and Z_m is the information measured, which is correlated to X_A.

Once the integration is calculated, we know the number of bits to remove by privacy amplification, namely $l \times I(X; Z) + |M_{\text{ow}}| + l \times I_{\text{int}}$, where $I_{\text{int}} = I(\Psi; S + E|Z)$ is the cost of interactivity. Since this cost is a part of the reconciliation, we define the net reconciliation efficiency by

$$\eta_{\text{net}} = \frac{H(\Psi(X)) - |M_{\text{ow}}|/l - I_{\text{int}}}{I(X; Y)}.$$

The results of the secret-key distillation are displayed in Table 11.3 for reverse reconciliation, which take into account the cost of interactivity, as in [77]. The reconciliation protocol is detailed in Section 9.3.3.

Table 11.3. *Results in the case of the realistic approach and reverse reconciliation. The evaluation of Eve's information due to the reconciliation is split into two parts: the bits sent one way $|M_{ow}|/l$ and the interactivity cost I_{int}. See also Table 9.2.*

G_{line}	$H(\Psi(X))$	I_{BE}	$|M_{\text{ow}}|/l$	I_{int}	η_{net}	Secret key	(kbps)
0.0 dB	4.63	0.00	2.50	0.000	89.0%	2.13	1700
1.0 dB	4.48	1.23	2.56	0.039	86.9%	0.65	520
1.7 dB	4.33	1.30	2.64	0.082	83.2%	0.31	250
3.1 dB	4.70	1.20	3.32	0.092	77.8%	0.09	75

11.5.4 Using a one-way binary correction protocol

Another option to overcome the problem of interactive reconciliation is to use a one-way binary correction protocol as often as possible. In [136], Nguyen and coworkers investigate the use of turbo codes as a replacement for Cascade. Actually, Cascade is not replaced for all the slices. Even if we count the parities disclosed from both sides, it still performs better than turbo codes for low error rates ($e \leq 8 \times 10^{-3}$). We can thus use either turbo codes or Cascade, whichever performs better for a given slice.

Since now all the disclosed bits are counted, we do not need to assume that Eve uses an ECM anymore. She can use whatever attack she wants; since the ECM maximizes her mutual information, she cannot get more information than what is computed by Alice and Bob.

The results are displayed in Table 11.4. The secret key rates come close to those of [77], and are even better for the 3.1 dB losses.

Table 11.4. Results in the case of the realistic approach and reverse reconciliation. Turbo codes are used. All the disclosed bits are counted in $|M|$. See also Table 9.3.

G_{line}	$H(\Psi(X))$	I_{BE}	$|M|/l$	η	Secret key	(kbps)
0.0 dB	4.51	0.00	2.51	83.9%	2.01	1605
1.0 dB	4.28	1.23	2.49	82.7%	0.56	450
1.7 dB	4.05	1.30	2.49	80.7%	0.26	210
3.1 dB	4.69	1.20	3.40	78.3%	0.10	80

11.6 Conclusion

In this chapter, I described two QKD protocols, one using squeezed states and one using coherent states, which are much easier to produce in practice. I also described the implementation of the protocol with coherent states, along with the secret-key distillation of its key elements.

The results are valid in the limit of arbitrarily large block sizes and assume that Eve is limited to individual attacks. In the next chapter, I explain how to remove such limitations and show how secret-key distillation can be more tightly integrated to the QKD protocol.

12
Security analysis of quantum key distribution

So far, the knowledge of Eve has been modeled by a classical random variable. The aim of this chapter is first to discuss how the previous results apply to QKD, where Eve's action may not necessarily be classically described by the random variable Z. Then, we explain the equivalence between BB84 and so-called entanglement purification protocols as a tool to encompass general eavesdropping strategies. Finally, we apply this equivalence to the case of the coherent-state protocol GG02.

12.1 Eavesdropping strategies and secret-key distillation

In Chapters 10 and 11, I analyzed the security of the protocols with regard to individual eavesdropping strategies only. In this particular case, the result of Alice's, Bob's and Eve's measurements can be described classically and the results of Section 6.4 apply. Let us review other kinds of eavesdropping strategies and discuss how the concepts of Chapter 6 can be applied.

Note that the individual eavesdropping strategy, the simplest class of strategies, is still technologically very challenging to achieve today in an optimal way. Nevertheless, we do not want the security of quantum cryptography to rely only on technological barriers. By considering more general eavesdropping strategies, we make sure quantum cryptography lies on strong grounds.

Following Gisin *et al.* [64], we divide the possible eavesdropping strategies into three categories: *individual attacks*, *collective attacks* and *joint attacks*.

- An *individual attack* is a strategy in which Eve probes each quantum state independently and measures each probe independently as well. In this case, the result of Eve's measurements can be modeled as the (classical) random variable Z.

- A *collective attack* is a strategy in which Eve probes each quantum state independently but she can measure the whole set of probes jointly. Furthermore, Eve can wait until all the public communications are completed, so that her measurement can depend on the exchanged messages. In this case, the result of Eve's actions cannot be modeled as a random variable per key element since we do not know what her measurement strategy is, prior to secret-key distillation. Collective attacks include individual attacks.

- A *joint attack* is a strategy in which Eve can probe all quantum states jointly. This is the most general kind of attack on the quantum channel. Of course, as for collective attacks, Eve can make the measurement of her choice after all the public communications are completed.

In the above three categories, it is tacitly assumed that Eve attacks the quantum key distribution protocol only. But nothing prevents her from waiting for Alice and Bob to use the key, e.g., for encrypting a secret message, before measuring her probe. The knowledge of what the legitimate parties do with their key might help Eve in breaking the cryptographic primitive for which Alice and Bob use the key. For instance, she may want to determine directly the secret message being encrypted. This is the problem of *composability* of quantum key distribution with other cryptographic protocols. I will discuss this aspect in Section 12.2.6. For now, Eve's goal is to determine the key – if she can determine the key, she can then break the cryptographic primitive that relies on the distributed key.

With an individual attack, Eve can use a quantum memory so that she can wait for Alice's sifting information and measure the correct observable. She can also wait for Alice and Bob to complete their reconciliation and privacy amplification protocols, but she does not gain anything from doing this: the measurement results are identically distributed before and after the secret-key distillation (SKD). We can thus safely assume that Eve measures her probes before reconciliation and privacy amplification.

With collective (and joint) attacks, Eve is not restricted to measuring the observable that gives the most information for each key element. With privacy amplification, a bit B from Alice's and Bob's secret key will ultimately be a function of several key elements $B = f(X_1, X_2, \ldots, X_l)$. Since Eve is now allowed to make a joint measurement, she can make her measurement depend on f and measure her states jointly in such a way that she gains as much information as possible on B after privacy amplification. In these circumstances, we can no longer assume that Eve measures her probes before

SKD. On the contrary, we must assume that she waits until reconciliation and privacy amplification are completed.

The SKD with collective attacks has been studied by Devetak and Winter [53]. They found that the secret key rate is lower bounded by $I(X;Y) - I'_{\text{AE or BE}}$, where $I'_{\text{AE or BE}}$ is the Holevo information between X and Eve's probe states [53, 83, 157]. This result is very similar to Theorem 3 of Section 6.4, with classical mutual information replaced by Holevo information.

The difference between collective and joint attacks is in the way Eve can interact with the states sent by Alice. With collective attacks, Eve must interact with each state independently, whereas joint attacks allow her to interact with the l states using any quantum operation on the l states.

From the literature, two ways to approach joint attacks can be mentioned. One way is to analyze the equivalence with entanglement purification – this will be discussed in detail in Section 12.2 below. The other way is to estimate the size of Eve's quantum memory, that is, the number of qubits needed to store her probes. Eve's probes can be seen as a black box on which queries can be made; in this case, the queries are all the (joint) measurements compatible with quantum mechanics. This is the concept of selectable knowledge defined by König, Maurer and Renner in [98, 99].

An important result in [99] is the following. For l large enough, let us consider l independent quantum carriers jointly probed by Eve and stored in a quantum memory of $l q_E$ qubits. Then, the number of secret key bits Alice and Bob can distill approaches $I(X;Y) - q_E$ per key element. This result is again very similar to Theorem 3 of Section 6.4, with classical mutual information replaced by Eve's number of qubits in her quantum memory. Christandl, Renner and Ekert [42] detail how Alice and Bob can estimate Eve's quantum memory size.

12.2 Distillation derived from entanglement purification

As a way to approach SKD with joint eavesdropping strategies, I shall consider SKD protocols that are equivalent to entanglement purification protocols. By first formally describing a protocol in a fully-quantum setting, we can derive a realistic protocol whose SKD part encompasses joint eavesdropping strategies, that is, it does not rely on any assumptions regarding what Eve can do on the quantum channel.

12.2.1 A formal equivalence

Entanglement purification was introduced by Bennett and coworkers [16]. Starting from a number of mixed non-maximally entangled quantum states shared between distant observers (Alice and Bob), an entanglement purification protocol (EPP) is aimed at getting a smaller number of close-to-pure close-to-maximally entangled quantum states. The only operations allowed are local quantum operations and classical communications.

Let me briefly describe why maximally entangled states are interesting for QKD. Maximally entangled states (also called EPR pairs [56]), at least those we consider here, are of the form $|\phi^+\rangle = (|00\rangle + |11\rangle)/2^{1/2}$. Half of the state stays at Alice's location and the other half is sent to Bob's apparatus. If Alice and Bob are guaranteed that they share a state $|\phi^+\rangle$, they can generate a perfectly secret key bit by each measuring their half. First, the measurement yields the same result on Alice's and Bob's sides, as long as Alice and Bob measure in the same basis (e.g., $\{|0\rangle, |1\rangle\}$ or $\{|+\rangle, |-\rangle\}$). Then, the probabilities of the bits 0 and 1 are equal. And finally, the state $|\phi^+\rangle$ is pure, perfectly factored from the environment and thus uncorrelated to Eve's knowledge.

In a less perfect situation, if Alice and Bob share a state having fidelity $1 - 2^{-\delta}$ with $|\phi^+\rangle^{\otimes k}$, Eve's information on the key is upper-bounded as $2^{-\delta}(2k + \delta + 1/\ln 2) + 2^{O(-2\delta)}$ [108]. Obtaining close-to-1 fidelity EPR pairs is thus a guarantee of secrecy. Showing that a QKD protocol, together with its SKD protocol, is equivalent to an EPP provides a proof that the generated key bits are indeed secret even if Eve is allowed to use any eavesdropping strategy.

The strategy we will use in this section is depicted in Fig. 12.1. A *prepare-and-measure* QKD protocol – like BB84 or GG02 – works as in the steps 1–2–3 of Fig. 12.1. Alice randomly chooses a key element X_A and prepares a quantum state that encodes it, which she sends to Bob. In step 1–2, Bob measures the state and obtains the key element Y_B. Finally, Alice and Bob perform SKD (step 2–3) to obtain a secret key.

It is important to stress that the EPP we require here is only used in a formal way, in the scope of this security analysis. There is no need to actually implement it, as this would be virtually impossible using current technologies.

With entanglement purification, Alice would prepare an entangled state, of which she would keep a subsystem called **a** and send the subsystem called **b** to Bob. On the way, Eve might interact with the state, resulting in a tripartite state $\rho_{a,b,eve}$. In step I–II of Fig. 12.1, Alice and Bob would

12.2 Distillation derived from entanglement purification

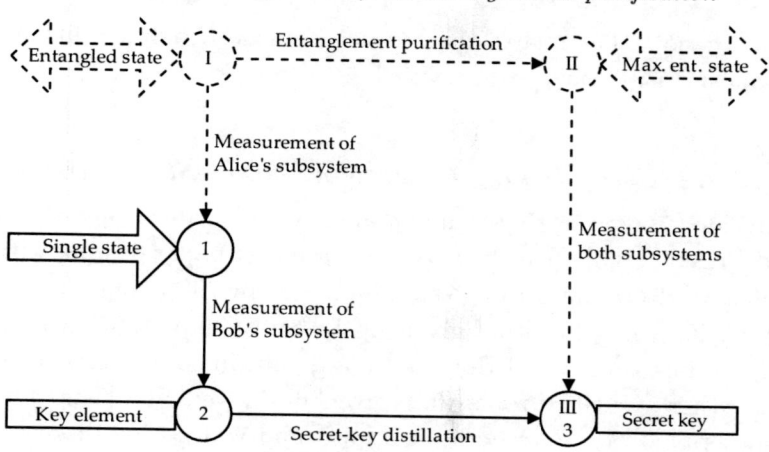

Fig. 12.1. The formal equivalence of a quantum key distribution protocol and its secret-key distillation with an entanglement purification protocol. The realistic protocol follows the path 1–2–3, whereas the formal protocol follows the path I–II–III. The dashed lines indicate the formal part, which does not have to be implemented in practice.

perform some EPP so as to obtain a state $\rho'_{a,b}$ close to $(|\phi^+\rangle\langle\phi^+|)^{\otimes k}$. Then in step II–III, Alice and Bob would measure their subsystem in the same basis and obtain a secret key as explained above.

Since the formal entangled state is measured by Alice and Bob after its interaction with Eve's probes, the state held by the eavesdropper does not depend on whether Alice and Bob perform their measurement before or after EPP. Stated otherwise, Eve's knowledge does not change if Alice and Bob follow the I–1–2 path or the II–III path.

What does make the paths I–1–2–3 and I–II–III equivalent and make them result in the identical situation 3=III as suggested on Fig. 12.1? First, the formal entangled state in I should be such that it is equivalent to Alice's modulation in 1 if she measures her subsystem **a**. Second, the EPP in step I–II has to be shown to be formally equivalent to the SKD in step 2–3.

Assuming that we can verify the two conditions stated above, the secrecy of the bits produced by the realistic protocol 1–2–3 is guaranteed by the ability of the formal EPP I–II to produce maximally entangled states.

Note that the path I-1-2-3 is used by the implementations of BB84 that rely on entangled photon pairs – see Section 10.2.1. With such implementations, the source produces entangled photon pairs (point I), half of which are immediately measured by Alice (step I–1) so as to determine what is being sent to Bob. The rest of the protocol, however, behaves like a regular prepare-and-measure one.

12.2.2 Using binary Calderbank–Shor–Steane codes

The equivalence between BB84 and entanglement purification was shown by Shor and Preskill [163]. It is, however, not the first paper dealing with BB84 and joint eavesdropping strategies. The paper by Shor and Preskill share some similarity with the proof given by Lo and Chau [108], although this last proof requires Alice and Bob to use a quantum computer, which is not realistic. Other security proofs were given by Mayers [126] and by Biham and coworkers [23]; these are quite complex and will not be discussed here.

In the case of BB84, the Calderbank–Shor–Steane (CSS) quantum codes [34, 168] can readily be used to establish the equivalence between an EPP and a QKD protocol [163]. Since we will use CSS codes as an ingredient for the entanglement purification and QKD protocols below, we will briefly review their properties. Before doing so, let us start again from point I.

Starting from the maximally entangled state

$$|\phi^+\rangle = 2^{-1/2}(|00\rangle_{ab} + |11\rangle_{ab}),$$

Alice keeps half of the state and sends the other half to Bob. This is the point I of Fig. 12.1. Let us now explain how it reduces to the point 1 of Fig. 12.1 when Alice measures her part. First, notice that $|\phi^+\rangle$ can be equivalently written as

$$|\phi^+\rangle = 2^{-1/2}(|++\rangle_{ab} + |--\rangle_{ab}).$$

The reduced system ρ_a gives a balanced mixture of $|0\rangle$ and $|1\rangle$ or equivalently a balanced mixture of $|+\rangle$ and $|-\rangle$, namely

$$\rho_a = \frac{1}{2}(|0\rangle\langle 0| + |1\rangle\langle 1|) = \frac{1}{2}(|+\rangle\langle +| + |-\rangle\langle -|).$$

Alice thus obtains balanced bits, as she would generate in BB84. Furthermore, the projection of system a onto $|i\rangle_a$ also projects b onto $|i\rangle_b$, for $i \in \{0, 1, +, -\}$. The identity of the state transmitted by Alice is thus perfectly correlated to her measurement result. Assuming that Alice randomly and uniformly chooses to measure a in $\{|0\rangle, |1\rangle\}$ or in $\{|+\rangle, |-\rangle\}$, it thus turns out to be as equivalent as if she transmits one of the states $\{|0\rangle, |1\rangle, |+\rangle, |-\rangle\}$ chosen randomly and uniformly as in BB84.

Let us come back again to the point I of Fig. 12.1. A number l of $|\phi^+\rangle$ states are produced and for each of these l states, the part b of $|\phi^+\rangle$ is sent to

12.2 Distillation derived from entanglement purification

Bob. Due to the channel losses or Eve's intervention, his part may undergo a bit error ($|\phi^+\rangle \to |\psi^+\rangle$), a phase error ($|\phi^+\rangle \to |\phi^-\rangle$) or both errors ($|\phi^+\rangle \to |\psi^-\rangle$), with $|\phi^-\rangle = 2^{-1/2}(|00\rangle - |11\rangle)$ and $|\psi^\pm\rangle = 2^{-1/2}(|01\rangle \pm |10\rangle)$. Given that not too many of these errors occur, Alice and Bob can obtain, from many instances of such a transmitted state, a smaller number of EPR pairs using only local operations and classical communications. One way to do this is to use CSS codes.

Let C^{bit} and C^{ph} be two binary error correcting codes of l bits (i.e., C^{bit} and C^{ph} are vector spaces of $\text{GF}(2)^l$). The parity check matrices of C^{bit} and C^{ph} are H^{bit} and H^{ph}, with sizes $l^{\text{bit}} \times l$ and $l^{\text{ph}} \times l$, respectively. They are chosen such that $\{0\} \subset C^{\text{ph}} \subset C^{\text{bit}} \subset \text{GF}(2)^l$. A CSS code is a k-dimensional subspace of $\mathcal{H}^l$, the Hilbert space of l qubits, with $k = \dim C^{\text{bit}} - \dim C^{\text{ph}} = l^{\text{ph}} - l^{\text{bit}}$ [34, 168].

The goal is for Alice and Bob to recover a state close to $|\phi^+\rangle^{\otimes k}$. They thus use the CSS code to correct the errors and improve the fidelity of their state with regard to $|\phi^+\rangle^{\otimes k}$ at the price of a decrease in the number of states obtained ($k \leq l$). The properties of the CSS codes are such that the code C^{bit} allows the correction of bit errors, while $C^{\text{ph}\perp}$ (the dual code of C^{ph}, with parity check matrix $H^{\text{ph}\perp}$ of size $l^{\text{ph}\perp} \times l$) allows the correction of phase errors. Furthermore, the correction of bit errors and phase errors can be performed independently.

The syndrome operators of CSS codes are of two kinds: the bit syndrome operators and the phase syndrome operators. The bit syndrome operators have the form $\mathbf{Z}^{h_1} \otimes \cdots \otimes \mathbf{Z}^{h_l}$, where $(h_1 \ldots h_l)$ is a row of H^{bit}, $h_i \in \text{GF}(2)$ and $\mathbf{Z}^0 = \mathbf{I}$ is the identity. Similarly, the phase syndrome operators have the form $\mathbf{X}^{h_1} \otimes \cdots \otimes \mathbf{X}^{h_l}$, where $(h_1 \ldots h_l)$ is a row of $H^{\text{ph}\perp}$. The eigenvalues of the syndrome operators are $\{\pm 1\}$. If the state belongs to the CSS subspace, the associated eigenvalues are $+1$ for all syndrome operators.

For purification, the idea is to use relative syndromes. When applying the CSS syndrome operators on either side of the state $|\phi^+\rangle^{\otimes l}$, there is no reason to measure only the eigenvalue $+1$; both values ± 1 can occur. However, the measured eigenvalues will be identical on both parts of $|\phi^+\rangle^{\otimes l}$. Hence, for purification, the legitimate parties only consider the relative bit syndromes $(\mathbf{Z_a} \otimes \mathbf{Z_b})^{h_1} \otimes \cdots \otimes (\mathbf{Z_a} \otimes \mathbf{Z_b})^{h_l}$ and the relative phase syndromes $(\mathbf{X_a} \otimes \mathbf{X_b})^{h_1} \otimes \cdots \otimes (\mathbf{X_a} \otimes \mathbf{X_b})^{h_l}$. For the state $|\phi^+\rangle^{\otimes l}$, the relative syndrome is always $+1$. Hence, purification with CSS codes consists in measuring the relative syndromes and apply corrections so as to ensure that the relative syndromes are all $+1$.

After the purification, the number of secret key bits is thus $k = \dim C^{\text{bit}} - \dim C^{\text{ph}}$, provided that C^{bit} (or $C^{\text{ph}\perp}$) is small enough to correct all the bit (or phase) errors. When considering asymptotically large block sizes, the CSS codes can produce

$$k = Rl \to l(1 - h(e^{\text{bit}}) - h(e^{\text{ph}})) = R^* l$$

EPR pairs or secret key bits, with e^{bit} (or e^{ph}) the bit (or phase) error rate and $h(p) = -p \log p - (1-p) \log(1-p)$ [163]. Here, $R = k/l$ indicates the rate obtained for a particular code and $R^* = 1 - h(e^{\text{bit}}) - h(e^{\text{ph}})$ is the asymptotically achievable rate. Note that the bit error rate e^{bit} determines the number of bits revealed by reconciliation (asymptotically $h(e^{\text{bit}})$), whereas the phase error rate e^{ph} determines the number of bits discarded by privacy amplification due to eavesdropping (asymptotically $h(e^{\text{ph}})$).

In the particular case of BB84, the secret key rate thus reads $I(X;Y) - h(e^{\text{ph}})$, which is similar to Theorem 3 of Section 6.4, with classical mutual information replaced by phase error correction information.

Bit and phase error rates estimation

Let us assume now that Alice sends either $|0\rangle$ or $|1\rangle$ to Bob (in the prepare-and-measure picture), or equivalently that Alice measures her part **a** in the $\{|0\rangle, |1\rangle\}$ basis. After sifting, Bob's bit values are determined by his measurement of **b** also in the $\{|0\rangle, |1\rangle\}$ basis. The bit error rate can thus be estimated on a sample set by Alice and Bob, who count the number of diverging key elements.

The phase error rate, however, cannot be estimated directly. For this, both Alice and Bob must use the $\{|+\rangle, |-\rangle\}$ basis. The number of erroneous results in this basis yields the number of phase errors in the other basis. Stated otherwise, the phase error rate of Alice and Bob using $\{|0\rangle, |1\rangle\}$ is equal to the bit error rate they would obtain using $\{|+\rangle, |-\rangle\}$. And vice versa, the phase error rate of Alice and Bob using $\{|+\rangle, |-\rangle\}$ is equal to the bit error rate they would obtain using $\{|0\rangle, |1\rangle\}$.

In the sequel, we can assume without loss of generality that Alice and Bob always use the $\{|0\rangle, |1\rangle\}$ basis to produce their key elements. The use of the other basis $\{|+\rangle, |-\rangle\}$ does not produce key elements but is needed to estimate the phase error rate and is essential to determine the eavesdropping level.

Eavesdropping implies phase errors

The interaction of Eve with the states sent by Alice can be such that no bit error occurs. But in this case, Eve cannot eavesdrop without introducing phase errors, as the bit and phase observables do not commute, $[\mathbf{X}, \mathbf{Z}] \neq 0$. Let me give some insight on this aspect using a toy model for eavesdropping.

12.2 Distillation derived from entanglement purification

Assume that the state $|\phi^+\rangle$ interacts with Eve's probes so as to make the 3-qubit state

$$|\Psi\rangle = \sum_{a,c \in \{0,1\}} f(a,c)|aac\rangle_{a,b,\text{eve}}.$$

We see that Bob's bit values will be strictly equal to Alice's, hence $e^{\text{bit}} = 0$. Let us further specify the state $|\Psi\rangle$ by correlating c to a: let c be equal to Alice's and Bob's value a affected by some error rate ϵ. More precisely, let $f(0,0) = f(1,1) = \sqrt{1/2 - \epsilon/2}$ and $f(0,1) = f(1,0) = \sqrt{\epsilon/2}$.

Let us first examine two extreme cases. When $\epsilon = 1/2$, Eve gets uncorrelated results, and one can check that $|\Psi\rangle = 2^{-1/2}|\phi^+\rangle_{\text{ab}}(|0\rangle_{\text{eve}} + |1\rangle_{\text{eve}})$. Tracing out Eve yields $\rho_{\text{ab}} = |\phi^+\rangle\langle\phi^+|$ and $e^{\text{ph}} = 0$. No further entanglement purification is necessary and all the states can be used to produce a secret key.

Then, let $\epsilon = 0$ so that Eve gets perfectly correlated results and $|\Psi\rangle = 2^{-1/2}(|000\rangle + |111\rangle)$. Tracing out Eve gives $\rho_{\text{ab}} = (|00\rangle\langle 00| + |11\rangle\langle 11|)/2$ and $e^{\text{ph}} = 1/2$. The phase error rate is too high, providing no secret key bits since $1 - h(e^{\text{ph}}) = 0$.

In the general case, one can determine the phase error rate by calculating $e^{\text{ph}} = \text{Tr}(\rho_{\text{ab}}(|\phi^-\rangle\langle\phi^-| + |\psi^-\rangle\langle\psi^-|))$. Since $e^{\text{bit}} = 0$ by construction, the expression can be simplified into $e^{\text{ph}} = \text{Tr}(\rho_{\text{ab}}|\phi^-\rangle\langle\phi^-|)$. By tracing out Eve, Alice's and Bob's subsystem reads

$$\rho_{\text{ab}} = \sum_{a,a'} C_{aaa'a'}|aa\rangle\langle a'a'|,$$

with $C_{aaa'a'} = \sum_c f(a,c)f^*(a',c)$. Combining the above expressions, one gets

$$e^{\text{ph}} = \frac{\sum_c |f(0,c) - f(1,c)|^2}{2} = \frac{(\sqrt{1-\epsilon} - \sqrt{\epsilon})^2}{2}.$$

The CSS code requires the removal of $h(e^{\text{ph}})$ bits to correct phase errors (in the entanglement purification picture) or for privacy amplification (in the prepare-and-measure picture). We assume that Eve measures her part in the computational basis and we compare this value to the order-2 Rényi entropy, as required in Theorem 1 of Section 6.3.1: $H_2(X|Z = e) = -\log((1-\epsilon)^2 + \epsilon^2)$ for $e \in \{0,1\}$. Since $h(e^{\text{ph}}) \geq H_2(X|Z = e)$, we verify that CSS codes give conservative results. Both curves are fairly close to each other, as one can see on Fig. 12.2.

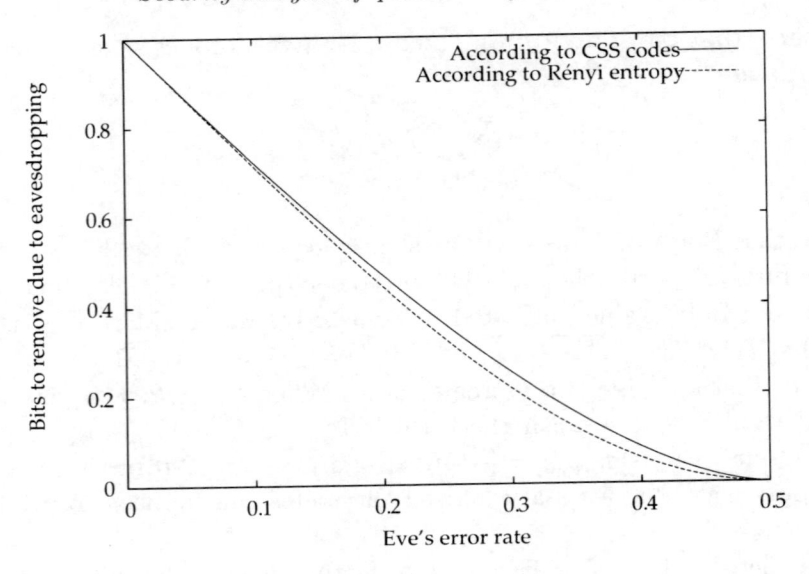

Fig. 12.2. The curves $h(e^{\mathrm{ph}})$ and $H_2(X|Z=e)$ as a function of ϵ.

12.2.3 Linking to practical secret-key distillation

Let me now discuss a little bit further the equivalence between the entanglement purification using CSS codes and the secret-key distillation for BB84.

For entanglement purification, Alice and Bob must compare their syndromes, both for bit errors and phase errors. The relative syndrome determines the correction that Bob must apply to align his qubits to Alice's.

As we have seen, the bit syndrome is determined by an operator of the form $\mathbf{Z}^{h_1} \otimes \cdots \otimes \mathbf{Z}^{h_l}$, for which the eigenvalues are $\{\pm 1\}$. For a state $|a_1 \ldots a_l\rangle$, the result of the syndrome measurement is $(-1)^{\sum_i h_i a_i}$ and is equivalent to the syndrome we would obtain classically, up to the ± 1 convention.

For relative syndromes, an operator has the form $(\mathbf{Z_a} \otimes \mathbf{Z_b})^{h_1} \otimes \cdots \otimes (\mathbf{Z_a} \otimes \mathbf{Z_b})^{h_l}$ and yields the result $(-1)^{\sum_i h_i(a_i+b_i)}$ on the state $|a_1 \ldots a_l\rangle_\mathbf{a} \otimes |b_1 \ldots b_l\rangle_\mathbf{b}$. This is again equivalent to what we would obtain classically.

For the phase errors, the situation is identical, with the $\mathbf{Z}$ operators replaced by $\mathbf{X}$ operators. By definition of the CSS codes, the phase and bit syndrome operators commute, allowing the CSS code to perform an independent correction of bit errors and phase errors. Let me thus split the discussion in two parts.

Bit error correction and reconciliation

For entanglement purification, Alice and Bob must calculate the relative syndrome, which determines the correction that Bob must apply to align his qubits to Alice's. Translating this into the BB84 protocol, one can show [163] that the relative syndrome for bit errors in the EPP is equal to the relative syndrome for bit errors that Alice and Bob would have reconciled in the BB84 protocol.

Referring to Fig. 12.1, the equivalence between the bit error correction part of I–II and the reconciliation part of 2–3 is straightforward. Any one-way binary error protocol such as turbo or LDPC codes discloses parities of the key elements, which can be expressed in the matrix H^{bit}. Bob's correction depends on the relative syndromes (parities). The code C^{bit} can thus be equivalently used both for reconciliation and for bit correction in the EPP.

For Cascade, Winnow or other interactive binary error correction protocols of Section 8.3, the equivalence is less straightforward and will be detailed in Section 12.2.4.

Phase error correction and privacy amplification

Phase errors of the EPP do not have such a direct equivalence in the BB84 protocol: the prepare-and-measure protocol works as if Alice measures her part of the state in the $\{|0\rangle, |1\rangle\}$ basis, thereby discarding information on the phase. However, one does not really need to correct the phase errors in the BB84 protocol. Instead, if $C^{\text{ph}\perp}$ is able to correct them in the EPP, the syndrome of C^{ph} in C^{bit} of Alice and Bob's bit string turns out to be a valid secret key in the prepare-and-measure protocol [163].

Since $C^{\text{ph}} \subset C^{\text{bit}}$, the matrix H^{ph} can be expressed using the same rows as H^{bit} and some additional rows:

$$H^{\text{ph}} = \begin{pmatrix} H^{\text{bit}} \\ H'^{\text{ph}} \end{pmatrix}. \qquad (12.1)$$

The secret key in the prepare-and-measure QKD protocol is given by $H'^{\text{ph}} x$, where $x \in \text{GF}(2)^l$ are the key elements. For the secret-key distillation to be equivalent to EPP using CSS codes, the privacy amplification must thus be linear. Furthermore, for H^{ph} to have full rank, the rows of H'^{ph} must be linearly independent of those of H^{bit}. Let us now discuss whether these requirements are compatible with privacy using universal families of hash functions, as described in Chapter 7.

The linearity requirement can be easily fulfilled using $\mathcal{H}_{\text{GF}(2^l) \to \{0,1\}^k}$, the family defined in Definition 10 of Section 7.2.3.

The linear dependence of rows in H^{bit} and in H'^{ph} has obvious consequences. The product $H^{\text{bit}}x$ contains publicly disclosed parities, whereas the product $H'^{\text{ph}}x$ is intended to produce the final secret key. Any linearly dependent row of H'^{ph} thus produces a bit that can be recovered by anyone who listens to the public channel during reconciliation.

The linear independence of a particular hash function with H^{bit} may not be easy to verify. The matrix H^{bit} may not have a particular structure. In the particular case of Cascade or turbo codes, the parities are generated from a pseudo-random permutation of the bits, hence making the H^{bit} matrix fairly general. Furthermore, the hash function is chosen randomly, preventing us from verifying this property of a particular hash function.

Intuitively, we may think that the matrix H'^{ph} of a randomly chosen hash function is unlikely to be linearly independent with a general reconciliation matrix H^{bit}. Yet, I show below that a randomly chosen linear hash function does a good job at producing a secret key, and that the number of linearly dependent rows is very small on average.

Theorem 13 *Let $\mathcal{H}$ be a universal family of linear hash functions with input set $\text{GF}(2)^l$ bits and output set $\text{GF}(2)^{l'^{\text{ph}}}$, with H'^{ph} the $l'^{\text{ph}} \times l$ matrix associated to a hash function of $\mathcal{H}$. Let H^{bit} be any fixed $l^{\text{bit}} \times l$ binary matrix and let H^{ph} be constructed by concatenating H^{bit} and H'^{ph} as in Eq. (12.1). Then,*

$$E[\text{rank } H^{\text{ph}}] \geq l^{\text{ph}} - 2^{l'^{\text{ph}} - (l - l^{\text{bit}})} / \ln 2,$$

with H'^{ph} chosen uniformly in $\mathcal{H}$.

Proof

Let X be a uniform random variable in $\text{GF}(2)^l$ such that $H^{\text{bit}}X$ is fixed. Thus, the order-2 Rényi entropy of X is $H_2(X) = l - l^{\text{bit}}$.

Let the uniform random choice of hash function be denoted by U. Thus $H(H'^{\text{ph}}X|U) = |\mathcal{H}|^{-1}\sum_u H(H'^{\text{ph}}X|U = u)$. For a given choice $U = u$ of hash function, H'^{ph} is fixed. Thus, $H(H'^{\text{ph}}X|U = u)$ is determined only by the number of rows in H'^{ph} that are linearly independent with those in H^{bit}, and $H(H'^{\text{ph}}X|U = u) = \text{rank } H^{\text{ph}} - l^{\text{bit}}$. Hence, $H(H'^{\text{ph}}X|U) = E[\text{rank } H^{\text{ph}}] - l^{\text{bit}}$.

Using Theorem 1 of Section 6.3.1, we obtain

$$H(H'^{\text{ph}}X|U) \geq l'^{\text{ph}} - 2^{l'^{\text{ph}} - (l - l^{\text{bit}})} / \ln 2,$$

hence the conclusion. □

Like Theorem 1, which does not guarantee that privacy amplification works for all choices of hash functions, Theorem 13 does not guarantee the full secrecy of the produced bits for all choices of hash functions. Yet, by choosing the matrix H'^{ph} small enough ($l'^{\text{ph}} < l - l^{\text{bit}}$), non-secret bits are produced with negligible probability.

Statistics and finite block sizes

Following the repetitive setting of Section 6.4, the number of bits to sacrifice for privacy amplification is evaluated using the mutual information. For finite block sizes, the actual number of bits to be sacrificed can be different. The convergence rate of the actual number of bits to the mutual information is difficult to estimate – e.g., see Section 11.5.1.

With protocols based on EPP, this problem becomes much simpler to deal with: all that matters is the statistical evaluation of the phase error rate. The actual phase error rate may be different from the one estimated. Yet, the statistical variance of the estimation with regard to the actual value is independent of the eavesdropper's action. Provided that the n evaluation samples are chosen randomly and uniformly among the $l + n$ transmitted states, Alice and Bob can infer an upper bound on the actual phase error rate; for instance, they can upper bound e^{ph} as $e^{\text{ph}} \leq \mu + s\sigma$, with μ their estimation, σ the standard deviation of the estimation and $s > 0$ a security parameter. Of course, Eve can be lucky and the actual phase error rate may be higher, but this happens only with a probability p_Q that exponentially decreases as s increases.

If we choose a hash function at random, there is also a probability that some key bits are correlated to the reconciliation information. Again, the probability of this event is independent of Eve's action and can be made as small as necessary.

So the equivalence of a QKD protocol and an EPP brings the advantage of being able to properly control finite block sizes.

I have thus shown that secret-key distillation equivalent to entanglement purification using CSS codes can be achieved using the existing techniques. I will now discuss the use of this technique in less ideal situations, such as when interactive reconciliation replaces syndromes of error correcting codes, and when Alice's source emits weak coherent states instead of single photons. Finally, I will show that the produced key is composable.

12.2.4 Interactive reconciliation

Entanglement purification with CSS codes essentially yields a protocol where the reconciliation is one way, i.e., the bit syndromes are sent one way from Claude to Dominique. What would be the entanglement purification equivalent of an interactive reconciliation protocol such as Cascade?

An important property of CSS codes is that the phase and bit syndrome operators commute. For interactive reconciliation, which bit syndromes are computed is not known beforehand but results from the interaction. Yet, we still need the bit syndrome operators to commute with the phase syndrome operators [69]. The problem comes from the bisection procedure of Cascade. At some point, the bisection narrows down to a single bit and its value is revealed. Translating this into entanglement purification, this means that the bit syndrome operator $\mathbf{Z}_j$, applying on the single qubit at position j, is measured at some point. However, any phase syndrome operator acting at least on the same qubit (i.e., including the factor $\mathbf{X}_j$) would not commute with $\mathbf{Z}_j$ since $[\mathbf{Z}_j, \mathbf{X}_j] \neq 0$. Hence, possible phase errors cannot be corrected on this particular qubit.

A solution was proposed by Lo [109]. The execution of Cascade does not depend on the particular values of the bits but on the differences encountered; or stated otherwise, Cascade depends only on the sum of Claude's and Dominique's bit strings. In the prepare-and-measure picture, the proposed method requires that Claude and Dominique encrypt the reconciliation messages using a key, for instance a fragment of the previously distributed key. Apparently, this may decrease the secret key rate of the previous block but since the reconciliation messages M are encrypted, privacy amplification does not have to take reconciliation messages into account. In fact, the cost of reconciliation is the same; it is counted towards the previous block instead of towards the current block.

Let $k_{1...r}$ be secret key bits known only to Claude and Dominique. For each syndrome bit $H_i x = \sum_{j=1...l} h_{ij} x_j$ (i.e., H_i is a row of H^{bit}) revealed by Claude, s/he encrypts it as $k_i \oplus H_i x$. Dominique always reveals the same function of bits as Claude. S/He also encrypts the result with the same key bit and sends $k_i + H_i y$. From the point of view of an observer, one can determine if there is a discrepancy between the bits of x and y, i.e., by calculating $k_i \oplus H_i x \oplus k_i \oplus H_i y = H_i(x \oplus y)$, but nothing can be known about the absolute value.

To understand why this works, let us go back to the entanglement purification picture. The key used to encrypt reconciliation messages is composed of pure EPR pairs $|\phi^+\rangle$ in registers $\mathsf{qa}_{1...r}$ and $\mathsf{qb}_{1...r}$. Instead of calculating the syndrome $\mathbf{H_a} = \mathbf{Z_a}^{h_1} \otimes \cdots \otimes \mathbf{Z_a}^{h_l}$ (or $\mathbf{H_b} = \mathbf{Z_b}^{h_1} \otimes \cdots \otimes \mathbf{Z_b}^{h_l}$), Claude (or Dominique) reveals $\mathbf{H_a} \otimes \mathbf{Z_{qa_i}}$ (or $\mathbf{H_b} \otimes \mathbf{Z_{qb_i}}$). Since Claude and Dominique are only interested in the relative value of the syndrome, they calculate $\mathbf{H_a} \otimes \mathbf{Z_{qa_i}} \otimes \mathbf{H_b} \otimes \mathbf{Z_{qb_i}}$. Since $\mathbf{qa}_i\mathbf{qb}_i$ contains the state $|\phi^+\rangle$, it gives the eigenvalue $+1$ for the operator $\mathbf{Z_{qa_i}} \otimes \mathbf{Z_{qb_i}}$ and the encrypted relative syndrome reduces to $\mathbf{H_a} \otimes \mathbf{H_b}$. Furthermore, it can be shown [109] that the state behaves as if $\mathbf{H_a} \otimes \mathbf{H_b}$ was measured, not the individual syndromes $\mathbf{H_a}$ and $\mathbf{H_b}$ alone. In particular, the bit value of a single qubit is not measured, only the relative value is.

12.2.5 Weak coherent states

So far, we have only looked at the case of perfect single-photon sources. If Alice emits weak coherent states, the equivalence with the EPP must be adapted.

As we saw, the multi-photon pulses can be attacked without inducing errors. The key elements carried by multi-photon pulses are called *tagged bits*, as if an adversary would tag the pulses as attackable. The ratio of tagged bits Δ can be found with the decoy state method described in Section 10.3.4. What remains is to determine the number of secret key bits that can be distilled as a function of Δ.

The equivalence of BB84 with an EPP based on CSS codes in the presence of tagged bits was studied by Gottesman, Lo, Lütkenhaus and Preskill [70]. The l qubits (in the entanglement purification picture) or key elements (in the prepare-and-measure picture) to process are of two kinds. First, $l\Delta$ of them are tagged, meaning that they were attacked without any visible errors. Second, the remaining $l(1-\Delta)$ pulses may suffer from possible phase errors due to Eve's intervention as in the single-photon case. Since Alice and Bob do not know if a particular pulse contains one or more photons, the perceived phase error rate is

$$e^{\text{ph}} = (1-\Delta)e^{\text{ph}}_{\text{untagged}} + \Delta e^{\text{ph}}_{\text{tagged}} \geq (1-\Delta)e^{\text{ph}}_{\text{untagged}}.$$

The reconciled key elements can be written as the sum of tagged and untagged key elements: $\Psi_{1\ldots l} = \Psi_{\text{tagged}} \oplus \Psi_{\text{untagged}}$. Since the hash function used for privacy amplification is linear, the resulting key is the sum of the tagged and untagged hashed key elements:

$$K_{1\ldots k} = H'^{\text{ph}}(\Psi_{\text{tagged}} \oplus \Psi_{\text{untagged}}) = K_{\text{tagged}} \oplus K_{\text{untagged}}.$$

In a sense, the tagged key bits are combined with the untagged bits using the one-time pad. Hence, the final key K is secret if K_{untagged} is secret, that is, if K_{untagged} is short enough. Here comes the important point: Alice and Bob have to choose the final key size $k = l^{\text{bit}} - l^{\text{ph}}$ so that they could distill pure EPR pairs from untagged qubits as if they knew which ones were untagged. So, even though they do not know which key elements are tagged or untagged, they can focus on the untagged qubits alone in the calculation of the key size.

Let us estimate the key size. From the $l\Delta$ tagged key elements, they cannot extract any secret information. Out of the remaining $l(1-\Delta)$ untagged key elements, they have to remove asymptotically $l(1-\Delta)h(e^{\text{ph}}_{\text{untagged}})$ bits for privacy amplification. In the worst case, all the phase errors are on untagged key elements, and thus $e^{\text{ph}}_{\text{untagged}} = e^{\text{ph}}/(1-\Delta)$. Therefore, the asymptotic secret key rate is

$$R^* = (1-\Delta) - h(e^{\text{bit}}) - (1-\Delta)h\left(\frac{e^{\text{ph}}}{1-\Delta}\right).$$

For realistic reconciliation, one has to replace $h(e^{\text{bit}})$ with $|M|/l$.

12.2.6 Composability

Informally, a QKD protocol is said to be *composable* if the key it produces is almost as good as if it were distributed with an ideal key distribution protocol. A cryptographic primitive, which is secure when used with an ideally secret key, must still be secure if used with a QKD-distributed key.

The QKD-distributed key must fulfill three conditions. First, the key must have a distribution that is close to uniform. Second, the event that the key reconciled by Alice and Bob is different must be rare. Finally, the mutual information of Eve with the produced keys must be small. Unsurprisingly, we expect these conditions to be met after reconciliation and privacy amplification.

What is especially important in the context of security of QKD against joint eavesdropping attacks, however, is that Eve's information on the key must be small even if she keeps quantum information until Alice and Bob actually use the key. If the eavesdropper keeps quantum information and delays her measurements until Alice and Bob reveal how they are going to use the key, can the chosen cryptographic primitive be broken? So the above three conditions must still be fulfilled in the presence of an eavesdropper with unlimited technology and no limit besides the laws of quantum mechanics.

Let K_A and K_B be the random variable containing the secret key as produced by a QKD protocol for Alice and Bob, respectively. We assign different variables to Alice's and Bob's keys because there is a non-zero probability that reconciliation does not correct all the errors. Note that the produced keys do not have a fixed length but can be any arbitrary string of bits, $K_A, K_B \in \{0,1\}^*$. The length of the keys, however, is publicly discussed by Alice and Bob over the public classical authenticated channel, so they must be equal: $|K_A| = |K_B| = |K|$. Depending on the quality of the quantum channel (and thus on the amount of eavesdropping), the final key size may vary. Also, Alice and Bob may decide to abort the protocol, in which case $|K| = 0$.

For QKD protocols based on EPP, the good news is that the produced key is composable if the purified state $\rho'_{a,b}$ is close to $|\phi^+\rangle^{\otimes |K|}\langle\phi^+|$. More precisely it is shown in [8] that a sufficient condition for composablility is

$$\sum_k \Pr[|K|=k](1 - F(|\phi^+\rangle^{\otimes k}\langle\phi^+|, \rho'_{a,b})) \leq \epsilon,$$

where $\epsilon > 0$ is a small constant that can be chosen as small as desired by the legitimate parties. As a convention, $F(|\phi^+\rangle^{\otimes 0}\langle\phi^+|, \rho'_{a,b}) = 1$, since Eve can acquire no advantage from an aborted protocol. For the cases where $|K| > 0$, the use of CSS codes with appropriate size can make the fidelity of the purified state $\rho'_{a,b}$ with the Bell state $|\phi^+\rangle^{\otimes k}\langle\phi^+|$ as close to 1 as desired [163].

12.3 Application to the GG02 protocol

In this section, we apply the formal equivalence between an EPP and the GG02 coherent-state protocol. Before we do so, let us mention some important studies of eavesdropping strategies against GG02.

GG02 is analyzed with regard to Gaussian individual attacks in [74, 77] and overviewed in Section 11.3.2. Grosshans and Cerf found the individual Gaussian attacks to be optimal in the more general class of non-Gaussian finite-width attacks [79]. A finite-width attack is such that, for some integer w, Eve interacts coherently with groups of up to w states at a time. Her interaction and measurement may not cross the borders of the groups of w states. This result is important in the sense that the secret key rates of Section 11.3.2 are also valid for this larger class of eavesdropping strategies, without the need for Alice and Bob to perform secret-key distillation differently.

In [89, 175], the equivalence of GG02 with entanglement purification is shown, thereby describing how to take joint eavesdropping strategies into account; this is the approach I wish to describe. Grosshans, Navascués and Acín [80, 133] calculated the secret key rates for the cases of collective eavesdropping and joint eavesdropping. These results will also be useful in this section.

In contrast to BB84 and its formal equivalence to CSS codes, the modulation of coherent states in GG02 is continuous, and therefore produces continuous key elements from which to extract a secret key. In this section, we generalize the sliced error correction (SEC) as an EPP. This way, we can show its equivalence with GG02 when complemented by the SEC. Owing to its generality, the asymptotic efficiency of this EPP develops, to some extent, from the asymptotic efficiency of the classical reconciliation protocol.

12.3.1 Rationale

Before diving into the technical details of the analysis below, let me overview the approach and explain the design choices.

Generalizing sliced error correction

On the one hand, the CSS codes provide a nice framework to analyze BB84 with regard to joint eavesdropping strategies. Entanglement purification with CSS codes is shown to be equivalent to BB84, together with appropriate binary reconciliation and privacy amplification protocols. On the other hand, GG02 also uses binary reconciliation protocols to produce secret key bits. More precisely, the SEC converts the key elements into bits, each called a slice, and thus comes down to the sequential use of a binary correction protocol – see Section 9.1. The next step is obviously to connect these two ideas, i.e., to generalize sliced error correction by using the CSS framework for binary correction.

The analysis below is centered around the generalization of sliced error correction as an EPP – see Section 12.3.4. The bits produced by the slices on Claude's side and by the slice estimators on Dominique's side become qubit pairs. In the prepare-and-measure picture, the bit strings are corrected using some linear binary error correction protocol; in the entanglement purification picture, the qubit pairs are purified using CSS codes. In a sense, each slice must be seen as an instance of BB84. Since sliced error correction typically produces more than one bit per continuous variable, we thus obtain more than one qubit pair to purify per use of the quantum channel.

Note that multistage soft decoding (see Section 9.2) does not convert the key elements of both Alice and Bob into bits before performing the binary reconciliation; only Claude's key elements are converted into bits. Since the CSS codes assume that one has qubits on both sides, we focus on sliced error correction instead of multistage soft decoding.

12.3 Application to the GG02 protocol

For the entanglement purification picture of BB84, one starts with l instances of $|\phi^+\rangle$ (point I of Fig. 12.1) and uses CSS codes to produce $k \leq l$ instances of $|\phi^+\rangle$ (point II). For GG02, we must use an entangled state such that it reduces to Alice's modulation when she measures her part (step I–1). As for BB84, we wish to produce instances of $|\phi^+\rangle$, but unlike as BB84, we start the purification with another initial state $|\Psi\rangle$, which is defined in Eq. (12.2) below.

The state $|\Psi\rangle$ is completely ad hoc. It does not have a particular physical meaning: the main requirement is that it correctly reduces to Alice's modulation. For our purposes, another requirement is that it makes the classical key elements that Alice and Bob would obtain in the prepare-and-measure picture appear clearly. For example, it uses the representation in the position basis so as to show the distribution of the homodyne detection results Bob would obtain.

The generalization of SEC as an EPP requires the description of the operations using unitary transformations. In particular, the operations must be reversible. In the classical sliced error correction, continuous key elements are converted into bits, disregarding a continuous part of the variable. Our first step is thus to describe the sliced error correction using reversible mappings – see Section 12.3.4.

The quantum generalization of sliced error correction must also be carefully designed so as to preserve phase information. As discussed in Section 12.2.2 above, the phase error rate must be kept as small as possible, as it otherwise reduces the number of maximally entangled states or secret key bits produced.

How to estimate the phase error rate

Unlike bit information, phase information is not directly accessible in the prepare-and-measure picture. Yet, phase error rate estimation is crucial for privacy amplification. For BB84, the symmetry between the $\{|0\rangle, |1\rangle\}$ and the $\{|+\rangle, |-\rangle\}$ bases allows one to easily estimate the phase error rate in one basis by estimating the bit error rate in the other basis. In the full generality of the quantum sliced error correction, such a symmetry property does not exist and the phase error rate must be determined otherwise; we use quantum tomography instead – see Section 12.3.3.

To understand the need for tomography, let us mention the squeezed-state protocol of Gottesman and Preskill [68] (GP01). The modulation in their protocol is identical to the Gaussian modulation of squeezed states of Section 11.2: The information is encoded in squeezed states displaced in either $\mathbf{x}$ or $\mathbf{p}$. Instead of using continuous key elements, however, they use the framework of CSS codes and encode qubit pairs in a formal entangled state. The protocol GP01 uses a simple qubit encoding scheme, with two rules (displacement in either **x** or **p**), and the symmetry of the encoding is such that a phase error for one rule corresponds to a bit error for the other rule. Hence, the estimation of the phase error rate is very simple, as for BB84.

Yet, the encoding in GP01 either is such that $R^* = 1 - h(e^{\text{bit}}) - h(e^{\text{ph}}) > 0$ or prevents one from producing any secret key bits. It is shown in [68] that GP01 must use squeezed states and that replacing them with coherent states would not work ($R^* = 0$). Thus, if we wish to use coherent states and produce a non-zero secret key rate, we may not rely on GP01's encoding symmetry properties to estimate the phase error rate. The alternative would be either to find another encoding with appropriate symmetry, which seems quite difficult, or to estimate the phase error rate by some other means and, in particular, using tomography.

Implementability and efficiency

We wish to show that the modulation of coherent states can be used to construct a QKD protocol based on entanglement purification and thus resistant to joint attacks. We try to keep the modulation of coherent states as it is, with as few changes to the physical implementation as possible. Most changes will be on the side of secret-key distillation, or at least on the way we approach it. Yet, we do not want merely to evaluate the secret key rate under various circumstances. It is important that the techniques used are implementable and not too far from the existing ones.

Although some changes are made to the GG02 protocol, we also wish to show that under reasonable circumstances, the protocol is still efficient and produces a high secret key rate. We will do this when we analyze the case of the attenuation channel and the asymptotic behavior – see Sections 12.3.5 and 12.3.6.

Unfortunately, the evaluation of the phase error rate is done with quantum tomography, which requires a lot of estimation samples, and this remains a practical issue.

12.3.2 Overview of the protocol

Let us define the ad hoc state (point I of Fig. 12.1) from which to start the purification in the case of the GG02 protocol. The formal state that Alice prepares must reduce to the proper modulation when Alice measures her

12.3 Application to the GG02 protocol

part. We define the formal state as:

$$|\Psi\rangle = \int \mathrm{d}x \mathrm{d}p \sqrt{p_{N(0,\Sigma_A \sqrt{N_0}) \times N(0,\Sigma_A \sqrt{N_0})}(x,p)} |x\rangle_{\mathsf{a}_1} \otimes |p\rangle_{\mathsf{a}_2} \otimes |x+ip\rangle_{\mathsf{b}}. \tag{12.2}$$

The kets $|x\rangle$, $|p\rangle$, $|x+ip\rangle$ are shorthand notations for, respectively, an **x**-quadrature eigenstate with eigenvalue x, a **p**-quadrature eigenstate with eigenvalue p and a coherent state whose **x** mean value equals x and whose **p** mean value equals p. The subscripts a_1, a_2 (or b) denote that the system is lying on Alice's side (or Bob's side).

The state in Eq. (12.2) does not have a direct physical meaning. In particular, the systems a_1 and a_2 must be understood as classical pointers, e.g., resulting from the (formal) homodyne detection measurement of an EPR state as studied in [78].

In the entanglement purification picture, the b part of the system is sent to Bob (and possibly attacked by Eve) and the $\mathsf{a}_1\mathsf{a}_2$ part stays at Alice's station. Notice that the measurement of both $\mathsf{x}_{\mathsf{a}_1}$ and $\mathsf{p}_{\mathsf{a}_2}$, giving $X_{A,1}$ and $X_{A,2}$ as a result, projects the subsystem b onto a coherent state of center $X_{A,1}+iX_{A,2}$. In this case, $X_{A,1}$ and $X_{A,2}$ are two independent Gaussian random variables, as in Section 11.3. Furthermore, the b subsystem, obtained by tracing out a_1 and a_2, yields the state Eq. (11.6). The state $|\Psi\rangle$ thus correctly reduces to the Gaussian modulation of coherent states, as in step I–1 of Fig. 12.1.

Let me now describe the EPP, which reduces to the prepare-and-measure QKD protocol described just afterwards.

- Alice creates $l+n$ copies of the state $|\Psi\rangle$, of which she sends the b part to Bob.
- Bob acknowledges reception of the states.
- Out of the $l+n$ states, n will serve for estimation purposes. These states are chosen randomly and uniformly by Alice, who informs Bob about their positions.
- For the remaining l states, Alice and Bob perform entanglement purification, so as to produce $k = Rl$ ($0 \leq R \leq 1$) states very close to $|\phi^+\rangle$. Measured in the computational bases, the produced states yield k secret bits on both Alice's and Bob's sides.

The details of the EPP, which uses CSS codes as an ingredient, are given in Section 12.3.4, while the estimation is detailed in Section 12.3.3.

By virtually measuring the $\mathsf{a}_1\mathsf{a}_2$ part of the state $|\Psi\rangle$, the protocol above reduces to the following one.

- Alice modulates $l+n$ coherent states $|x+ip\rangle$ that she sends to Bob. The values of x and p both follow an independent Gaussian distribution with zero mean and variance $\Sigma_A^2 N_0$.

- Bob acknowledges reception of the states.
- Out of the $l+n$ states, n will serve for estimation purposes. These states are chosen randomly and uniformly by Alice, who informs Bob about their positions.
- For the remaining l states, Bob measures $\mathbf{x}$. Alice and Bob perform secret-key distillation (reconciliation and privacy amplification), so as to produce $k = Rl$ secret bits.

The reconciliation and privacy amplification procedures are based on classical error correcting codes, which are derived from the CSS codes used in the formal EPP.

12.3.3 Error rates estimation using tomography

In QKD protocols derived from EPP, an important step is to show how one can infer the bit and phase error rates of the samples that compose the key. A fraction of the samples sent by Alice to Bob is sacrificed so as to serve as test samples. By randomly choosing them within the stream of data, they are statistically representative of the whole stream.

In [68, 163], one can simply make measurements and directly count the number of bit and phase errors from the results. This is possible since Bob's apparatus can measure both bit and phase values. In our case, however, it is not possible to directly measure phase errors. Yet some data post-processing can be applied on measurements so as to infer the number of phase errors in the stream of data.

The encoding of bits will be described in a further section. For the moment, the qubit pair system, which Alice and Bob will process using CSS codes, is not explicitly described. However, it is sufficient to describe the CSS codes in terms of the Pauli bit-flip and phase-flip operators of Alice's qubit system in $\mathbf{a}_1$, namely $\mathbf{Z_s}$ (phase flip) and $\mathbf{X_s}$ (bit flip), and of the Pauli operators in Bob's qubit system in $\mathbf{b}$, namely $\mathbf{Z_e}$ and $\mathbf{X_e}$. (The subscripts s and e stand for slice and estimator, respectively, to follow the convention of the following sections.) The bit errors are assumed to be easy to determine, that is, $\mathbf{Z_s}$ has a diagonal expansion in $|x\rangle_{\mathbf{a}_1}\langle x|$, and $\mathbf{Z_e}$ can be directly determined by a single homodyne detection measurement on $\mathbf{b}$. This ensures, in the derived prepare-and-measure QKD protocol, that Alice knows the bit value she sent, and Bob can determine the received bit value. A measurement of the observable $\mathbf{X_s I_{a_2} X_e}$ associated to the phase error rate, however, cannot be implemented by a single homodyne detection measurement on b. Therefore, we have to invoke quantum tomography with a quorum of operators [51] to get an estimate of the phase error rate.

Estimating phase errors in the average state

In the entanglement purification picture, let $\rho^{(n)}$ be the state of the n samples used for estimation of the phase error rate, i.e., n instances of the $\mathsf{a_1 a_2 b}$ system. To count the number of phase errors in a set of n samples, one needs to measure $\mathbf{O} = \mathbf{X_s I_{a_2} X_e}$ on the n samples and sum the results. This is equivalent to measuring $\mathbf{O}^{(n)} = \sum_i \mathbf{I}_{\mathsf{a_1 a_2 b}}^{\otimes i-1} \otimes \mathbf{X_s I_{a_2} X_e} \otimes \mathbf{I}_{\mathsf{a_1 a_2 b}}^{\otimes n-i}$. If the true phase error probability in the $n+l$ samples is e^{ph}, the error variance is $\sigma_1^2 = 2e^{\mathrm{ph}}(1-e^{\mathrm{ph}})/n$, and thus the probability of making an estimation error of more than Δ is asymptotically $\exp(-\Delta^2 n/4e^{\mathrm{ph}}(1-e^{\mathrm{ph}}))$ [68, 163]. It is easy to see that

$$\mathrm{Tr}(\mathbf{O}^{(n)}\rho^{(n)}) = n\,\mathrm{Tr}(\mathbf{O}\rho),$$

where $\rho = n^{-1}\sum_i \mathrm{Tr}_{\mathrm{All}\setminus\{i\}}(\rho^{(n)})$ is the density matrix of the average state measured. So, we can estimate the number of phase errors using the average state, even if the eavesdropper interacts jointly with all the states ($\rho^{\otimes n} \neq \rho^{(n)}$), in which case we say that the eavesdropping is joint.

If the measurement of $\mathbf{O} = \mathbf{X_s I_{a_2} X_e}$ cannot be made directly, one instead looks for a quorum of operators $\mathbf{Q}_\lambda$ such that $\mathbf{O} = \int \mathrm{d}\lambda o(\lambda) \mathbf{Q}_\lambda$; estimating $\langle \mathbf{O} \rangle$ comes down to measuring several times $\mathbf{Q}_\lambda$ for values of λ chosen randomly and independently from each other, and averaging the results weighted by $o(\lambda)$: $\mathbf{O} \approx \sum_i o(\lambda_i) \mathbf{Q}_{\lambda_i}$ [51]. If the values of λ are chosen independently from the sample index on which $\mathbf{Q}_\lambda$ is applied, we get unbiased results, as $\mathrm{Tr}(\mathbf{O}\rho) = E_\lambda[\mathrm{Tr}(\mathbf{Q}_\lambda \rho)]$, with E the expectation. Of course, the estimation of $\mathrm{Tr}(\mathbf{O}\rho)$ with a quorum cannot be perfect and results in an estimation variance σ_2^2. The variance of the estimated $\langle \mathbf{O} \rangle$ must increase by this amount, and the resulting total variance is $\sigma^2 = \sigma_1^2 + \sigma_2^2$.

Estimating phase errors using coherent states and homodyne detection

We now explain how the phase error rate can be estimated, in principle, using coherent states modulated in both quadratures and homodyne detection in all quadratures.

It is clear that the analytical expression of the average state ρ gives the knowledge of $\langle \mathbf{O} \rangle$. Let $\rho_0 = |\Psi\rangle\langle\Psi|$ be the state that Alice and Bob would share if the transmission were perfect. Since the a part of the system stays at Alice's station, we only need to learn about how the b part of the system is affected.

To model the quantum channel between Alice and Bob, let T be a mapping that maps input states ρ_{in} onto output states $\rho_{\text{out}} = T(\rho_{\text{in}})$. To model a quantum transmission channel properly, T must be linear and positive, i.e., $T(\rho_{\text{in}})$ must be positive whenever ρ_{in} is positive. It must also be completely positive (CP), that is, if we adjoin an arbitrary auxiliary system aux in a t-dimensional Hilbert space $\mathcal{H}^t$ (for any t), $T \otimes \text{Id}_{\text{aux}}$ must also be positive, where Id_{aux} is the identity in the auxiliary system. We call T a CP map.

In the prepare-and-measure picture, let T be the CP map that maps the states sent by Alice onto the states received by Bob in the system b, $(\text{Id}_{a_1 a_2} \otimes T_{\text{b}})(\rho_0) = \rho$. In particular, let the coherent state $|x+ip\rangle\langle x+ip|$ be mapped onto $\rho_T(x+ip)$ and the (pseudo-)position state $|x\rangle\langle x'|$ be mapped onto $\rho_T(x, x')$. The functions $\rho_T(x+ip)$ and $\rho_T(x, x')$ are related by the following identity:

$$\rho_T(x+ip) \propto \int dx' dx'' e^{-(x'-x)^2/4N_0 - (x''-x)^2/4N_0} e^{i(x'-x'')p/2N_0} \rho_T(x', x''),$$

with N_0 the variance of the vacuum fluctuations. By setting $D = x' - x''$ and $S = x' + x'' - 2x$, we get:

$$\rho_T(x+ip) \propto \int dD\,dS\, e^{-S^2/8N_0 - D^2/8N_0 + iDp/2N_0} \rho_T(x+S+D, x+S-D), \tag{12.3}$$

which shows that $\rho_T(x, x')$ is integrated with an invertible kernel (Gaussian convolution in S, multiplication by $e^{-D^2/8N_0}$, and Fourier transform in D). So in principle, any different CP map $T' \neq T$ implies a different effect on coherent states, $\rho_T(x+ip) \neq \rho_{T'}(x+ip)$. The modulation of coherent states in both quadratures is thus crucial for this implication to be possible.

By inspecting Eq. (12.3), it seems that because of the factors $e^{-S^2/8N_0}$ and $e^{-D^2/8N_0}$, two different CP-maps T and T' may make $\rho_T(x+ip)$ and $\rho_{T'}(x+ip)$ only vanishingly different. It thus seems unlikely that Eq. (12.3) should allow us to extract the coefficients $\rho_T(x+S+D, x+S-D)$. However, assuming that T depends only on a finite number of parameters, a variation of these parameters will induce a measurable variation of $\rho_T(x+ip)$. I will now discuss why it is reasonable to make such an assumption.

Because of the finite variance of the modulation of coherent states, the probability of emission of a large number of photons vanishes – this intuitively indicates that we only need to consider the description of T for a bounded number of emitted photons. More precisely, one can consider the emission of d joint copies of the state $\rho_{0b} = \text{Tr}_a(\rho_0)$. For d sufficiently large, $\rho_{0b}^{\otimes d}$ can be represented in the typical subspace $\Gamma_\delta(\rho_{0b})$ of dimension not greater than $2^{d(H(\rho_{0b})+\delta)}$, for any $\delta > 0$ [157], where $H(\rho)$ is the von Neumann entropy of a state ρ. The probability mass of $\rho_{0b}^{\otimes d}$ outside the typical subspace can be made arbitrarily small and does not depend on the eavesdropping strategy. This means that the support for the input of T has finite dimensionality, up to an arbitrarily small deviation.

The number of photons received by Bob can also be upper bounded. Alice and Bob can first assume that no more than $n_{\max}$ photons are received. This fact may depend on a malicious eavesdropper, so Bob has to carry out hypothesis testing. The test comes down to estimating $\langle \Pi \rangle$ with $\Pi = \sum_{n > n_{\max}} |n\rangle \langle n|$ in the Fock basis. If the threshold is well chosen so that $n > n_{\max}$ never occurs in practice, we can apply the central limit theorem and upper bound the probability that $\langle \Pi \rangle > \epsilon$ for any chosen $\epsilon > 0$. The positivity of the density matrices implies that the off-diagonal coefficients are also bounded. We can thus now express $\rho_T(x + ip)$ as $\rho_T(x + ip) = \sum_{n,n' \leq n_{\max}} \rho_T(x + ip, n, n') |n\rangle \langle n'|$. Note that the test can be implemented either by explicitly measuring the intensity of the beam (therefore requiring an additional photodetector) or by exploiting the correlation between the high intensity of the beam and the high absolute values obtained when making homodyne detection measurements in all directions.

Finally, the coefficient of $|n\rangle \langle n'|$ can be estimated with arbitrarily small statistical error using homodyne detection in all directions [50, 51]. This is achieved by considering the quorum of operators $(\mathbf{x}_\theta)_{0 \leq \theta < 2\pi}$, where $\mathbf{x}_\theta = \cos\theta\, \mathbf{x} + \sin\theta\, \mathbf{p}$ denotes the amplitude of the quadrature in direction θ. Considering a finite combination of arbitrarily small statistical errors on parameters also gives arbitrarily small overall statistical error on the phase error rate.

12.3.4 Encoding of multiple qubits in a continuous state

Reconciliation and privacy amplification are integral parts of the prepare-and-measure protocols derived from entanglement purification protocols. In our case, we wish to derive a prepare-and-measure protocol with sliced error correction (SEC) as reconciliation – see Section 9.1. We therefore need to construct an entanglement purification procedure that reduces to SEC when the corresponding prepare-and-measure protocol is derived.

Sliced error correction with invertible mappings

We here recall the main principles of SEC in a form that is slightly different from the presentation in Section 9.1. To suit our needs, we need to describe SEC in terms of invertible functions giving the slices and the estimators – the invertibility property will be required when we generalize SEC to entanglement purification. Also, two parameters are fixed here: binary error correction is operated by sending syndromes of classical linear error-correcting codes (ECC), and we momentarily restrict ourselves to the case of one-dimensional real values $X, Y \in \mathbf{R}$.

Suppose Claude and Dominique have l independent outcomes of X and Y, respectively, denoted by $x_{1...l}$ and $y_{1...l}$, from which they intend to extract common bits.

First, Claude converts each of her variables X into m bits and thereby defines m binary functions: $S_1(x), \ldots, S_m(x)$. To make the mapping invertible, she also defines a function $\bar{S}(x)$ such that mapping from the domain $\mathbf{R}$ to the vector $(\bar{S}(x), S_{1...m}(x))$ is bijective. As a convention, the range of $\bar{S}(x)$ is $[0; 1]$. We thus define the mapping $\mathcal{S}$ as

$$\mathcal{S} : \mathbf{R} \to [0; 1] \times \{0, 1\}^m : x \to (\bar{S}(x), S_{1...m}(x)).$$

As in Section 9.3.2, the functions $S_i(x)$ implicitly cut the real line into intervals. With the invertible mapping $\mathcal{S}$, we add the function $\bar{S}(x)$, which indicates where to find x within a given interval.

Then, we can assemble the bits produced by the l outcomes $x_1 \ldots x_l$ into m l-bit vectors. An ECC, upon which Claude and Dominique agreed, is associated with each bit vector ("slice") $S_i(x_{1...l}) = (S_i(x_1), \ldots, S_i(x_l))$. To proceed with the correction, Claude sends the syndrome $\xi_i^{\text{bit}} = H_i^{\text{bit}} S_i(x_{1...l})$ to Dominique over the public classical authenticated channel, where H_i^{bit} is the $l_i^{\text{bit}} \times l$ parity check matrix of the ECC associated to slice i. Claude also sends $\bar{S}(x_{1...l})$.

Dominique would like to recover $S_{1...m}(X_{1...l})$ from his/her knowledge of $Y_{1...l}$, $\xi_{1...m}^{\text{bit}}$ and $\bar{S}(X_{1...l})$. S/He also converts each of his/her outcomes $y_{1...l}$ into m bits, using the slice estimator functions $E_i(y, \bar{S}(x), S_{1...i-1}(x))$. Referring to Section 9.1, Dominique corrects the slices sequentially, thereby acquiring the knowledge of $S_{1...i-1}(x)$ before evaluating $E_i(y, \bar{S}(x), S_{1...i-1}(x))$. In addition to the presentation in Section 9.1, the estimator can also depend on $\bar{S}(x)$; one can think of $\bar{S}(x)$ as the fully disclosed slices – see Section 9.1.2.

Note that the estimators can also be written as jointly working on l samples at once: $E_i(y_{1...l}, \bar{S}(x_{1...l}), \xi_1^{\text{bit}}, \ldots, \xi_{i-1}^{\text{bit}})$, but we will preferably use the previous notation for its simplicity since, besides the ECC decoding, all the operations are done on each variable X or Y independently.

Like $\bar{S}(x)$, we also need a supplementary function $\bar{E}$ to ensure that the process on Dominique's side is described using bijective functions:

$$\bar{E}(y, \bar{S}(x), S_{1...m}(x))$$

(or jointly $\bar{E}(y_{1...l}, \bar{S}(x_{1...l}), \xi_1^{\text{bit}}, \ldots, \xi_m^{\text{bit}})$). As the knowledge of $S_{1...m}(x)$ is required, this function is evaluated after all the slices are corrected. As a convention, the range of $\bar{E}$ is $[0;1]$. $\bar{E}$ is chosen so that the mapping $\mathcal{E}$ defined below is invertible,

$$\mathcal{E} : [0;1] \times \{0,1\}^m \times \mathbf{R} \to [0;1] \times \{0,1\}^m \times \{0,1\}^m \times [0;1] :$$
$$(\bar{s}, s'_{1\ldots m}, y) \to (\bar{s}, s'_{1\ldots m}, E_1(y, \bar{s}), \ldots, E_m(y, \bar{s}, s'_{1\ldots m-1}), \bar{E}(y, \bar{s}, s'_{1\ldots m})).$$

Like $\mathcal{S}$, the functions $E_{1\ldots m}$ of $\mathcal{E}$ cut the real line into intervals. However, these intervals are adapted as a function of the information sent by Claude, so as to estimate Claude's bits more reliably. As for $\bar{S}$, the function $\bar{E}$ indicates where to find y within an interval.

The mapping $\mathcal{S}$ summarizes Claude's process of conversion of his/her real variable X into m bits (plus a continuous component). The mapping $\mathcal{E}$ represents the bits (and a continuous component) produced by Dominique from the real variable Y and the knowledge of $\bar{S}(X)$ and of the syndromes $\xi^{\text{bit}}_{1\ldots m}$. The bits produced by the functions E_i are not yet corrected by the ECC, even though they take as input the corrected values of the previous slices $S_j(X)$, $j < i$. The description of the mapping $\mathcal{E}$ with the bits prior to ECC correction allows us to express the bit error rate between Claude's slices and Dominique's estimators easily and, thereby, to deduce the size of the parity matrices of the ECCs needed for the binary correction. Simply, we define $e^{\text{bit}}_i = \Pr[S_i(X) \neq E_i(Y, \bar{S}(X), S_{1\ldots i-1}(X))]$. As the block size $l \to \infty$, there exist ECCs with size $l^{\text{bit}}_i \to lh(e^{\text{bit}}_i)$ and arbitrarily low probability of decoding error. The number of common (but not necessarily secret) bits produced by SEC is therefore asymptotically equal to $H(S_{1\ldots m}(X)) - \sum_{i=1}^{m} h(e^{\text{bit}}_i)$ per sample.

The generalization of the SEC to a quantum entanglement purification protocol is examined next.

Quantum sliced error correction

As the classical sliced error correction describes the reconciliation in step 2–3 of Fig. 12.1, I now describe its quantum generalization, aimed at defining the EPP of step I–II of Fig. 12.1. For our convenience, we will follow the direct reconciliation and stick to the Alice/Bob convention, as the link between the QKD protocol and the secret-key distillation is tight. We defer the treatment of reverse reconciliation until Section 12.3.6.

The purification uses a few quantum registers, which I will now list. Alice's system a_1 is split into m qubit systems $\mathsf{s}_{1\ldots m}$ and a continuous register $\bar{\mathsf{s}}$. On Bob's side, the system b is split into m qubit systems $\mathsf{e}_{1\ldots m}$ and a continuous register $\bar{\mathsf{e}}$. He also needs m qubit registers $\mathsf{s}'_{1\ldots m}$ for temporary storage. All these registers must, of course, be understood per exchanged sample: as Alice generates l copies of the state $|\Psi\rangle$, the legitimate parties use l instances of the registers listed above.

The usual bit-flip and phase-flip operators **X** and **Z**, respectively, can be defined as acting on a specific qubit register among the systems s_i and e_i. For example, $\mathbf{Z}_{s_i}$ is defined as acting on s_i only. These operators are used by Alice and Bob to construct the CSS codes that produce entangled qubits, which are in turn used to produce EPR pairs in the registers $s_i e_i$ for $i = 1 \ldots m$. Since each CSS code operates in its own register pair, the action of one does not interfere with the action of the other. It is thus possible to extract more than one EPR pair $|\phi^+\rangle$ per state $|\Psi\rangle$. If asymptotically efficient binary codes are used, the rate of EPR pairs produced is $R^* = \sum_i (1 - h(e_i^{\text{bit}}) - h(e_i^{\text{ph}}))$, where e_i^{bit} (or e_i^{ph}) indicates the bit error rate (or the phase error rate) [163].

The process that defines the content of the registers is described next.

The mappings $\mathcal{QS}$ and $\mathcal{QE}$

First, we define the unitary transformation $\mathcal{QS}$ by its application on the basis of quadrature eigenstates:

$$\mathcal{QS} : L^2(\mathbf{R}) \to L^2([0;1]) \otimes \mathcal{H}^{\otimes m} :$$
$$|x\rangle_{a_1} \to \sigma(x) |\bar{S}(x)\rangle_{\bar{s}} \otimes |S_1(x)\rangle_{s_1} \otimes \cdots \otimes |S_m(x)\rangle_{s_m}. \quad (12.4)$$

The states $|\bar{s}\rangle_{\bar{s}}$, $0 \leq \bar{s} \leq 1$, form an orthogonal basis of $L^2([0;1])$, $\sigma(x) = (d_x \bar{S})^{-1/2}(x)$ is a normalization function, and $|s_i\rangle_{s_i}$, $s_i \in \{0, 1\}$, denotes the canonical basis of $\mathcal{H}$, the Hilbert space of a qubit. As a convention, the system s_i is called slice i. The transformation $\mathcal{QS}$ is depicted in Fig. 12.3.

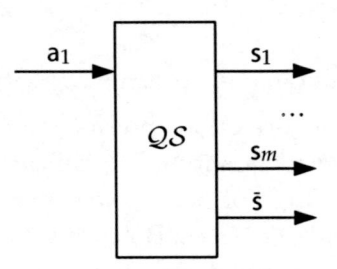

Fig. 12.3. Schematic description of $\mathcal{QS}$. Reprinted with permission from [175] © 2005 by the American Physical Society.

For each slice i, Alice and Bob agree on a CSS code, defined by its parity matrices H_i^{bit} for bit error correction and $H_i^{\text{ph}\perp}$ for phase error correction.

12.3 Application to the GG02 protocol

For the entanglement purification, let us assume that Alice computes the syndromes of the CSS code with a quantum circuit. For each slice, she produces l_i^{bit} qubits in the state $|\xi_i^{\text{bit}}\rangle$ and $l_i^{\text{ph}\perp}$ qubits in the state $|\xi_i^{\text{ph}}\rangle$ that she sends to Bob over a perfect quantum channel, so that the syndromes are received without any distortion. After reduction to a prepare-and-measure protocol, this perfect transmission is actually made over the public classical authenticated channel. Alice also sends the $\bar{s}$ system to Bob.

Then, the slice estimators are defined as the unitary transformation $\mathcal{QE}$,

$$\mathcal{QE} : L^2([0;1]) \otimes \mathcal{H}^{\otimes m} \otimes L^2(\mathbf{R}) \to L^2([0;1]) \otimes \mathcal{H}^{\otimes m} \otimes \mathcal{H}^{\otimes m} \otimes L^2([0;1]) :$$
$$|\bar{s}\rangle_{\bar{s}} |s'_{1...m}\rangle_{\mathsf{s}'_{1...m}} |y\rangle_{\mathsf{b}} \to$$
$$\epsilon(y, \bar{s}, s'_{1...m}) |\bar{s}\rangle_{\bar{s}} |s'_{1...m}\rangle_{\mathsf{s}'_{1...m}} \overset{m}{\underset{i=1}{\otimes}} |E_i(y, \bar{s}, s'_{1...i-1})\rangle_{\mathsf{e}_i} |\bar{E}(y, \bar{s}, s'_{1...m})\rangle_{\bar{\mathsf{e}}}, \quad (12.5)$$

where $\epsilon(y, \bar{s}, s'_{1...m}) = (\partial_y \bar{E})^{-1/2}(y, \bar{s}, s'_{1...m})$ is a normalization function; $|y\rangle_{\mathsf{b}}$ is a quadrature eigenstate with **x**-eigenvalue y; $|e_i\rangle_{\mathsf{e}_i}$, $e_i \in \{0, 1\}$, denotes the canonical basis of $\mathcal{H}$ and $|\bar{e}\rangle_{\bar{\mathsf{e}}}$, $0 \le \bar{e} \le 1$, form an orthogonal basis of $L^2([0;1])$. As the classical mapping $\mathcal{E}$ is invertible, $\mathcal{QE}$ is unitary with the appropriate normalization function ϵ. This mapping is defined to act on individual states, with the slice values $s'_{1...m}$ as input in the system $\mathsf{s}'_{1...m}$, whose purpose is actually to hold Bob's sequentially corrected bit values. The complete transformation jointly involving l systems would be fairly heavy to describe. Only the ECC correction needs to be described jointly, and assuming it is correctly sized (i.e., l_i^{bit} are large enough), Bob has enough information to reconstruct Alice's bit values. Let me now sketch how the system $\mathsf{s}'_{1...m}$ is constructed.

Assume that Bob first calculates, using a quantum circuit, the first slice estimator (classically: $E_1(Y, \bar{S}(X))$), which does not depend on any syndrome. That is, he applies the following mapping, defined on the bases of $\bar{\mathsf{s}}$ and b: $|\bar{s}\rangle_{\bar{s}} |y\rangle_{\mathsf{b}} \to |\bar{s}\rangle_{\bar{s}} |E_1(y, \bar{s})\rangle_{\mathsf{e}_1} |\bar{E}_1(y, \bar{s})\rangle_{\bar{\mathsf{e}}_1}$ (up to normalization), where the function $\bar{E}_1$ is needed only to make the mapping unitary. From the l qubits in the l systems e_1 and the syndrome sent by Alice $|\xi_1^{\text{bit}}\rangle$, there exists a quantum circuit that calculates the relative syndrome of Alice's and Bob's bits, that is a superposition of the classical quantities $\xi_1^{\text{bit}} \oplus H_1^{\text{bit}} E_1(X_{1...l})$. From this, a quantum circuit calculates the co-set leader of the syndrome, that is (a superposition of) the most probable difference vector between Alice's and Bob's qubits. An extra $l - l_1^{\text{bit}}$ blank qubits are needed for this operation; we assume they are all initialized to $|0\rangle$:

$$|H_1^{\mathsf{b}}(s_1^{(l)} \oplus e_1^{(l)})\rangle_{\mathsf{s}_1'^{(l_1^{\mathsf{b}})}} |0\rangle_{\mathsf{s}_1'^{(l-l_1^{\mathsf{b}})}} \to |s_1^{(l)} \oplus e_1^{(l)}\rangle_{\mathsf{s}_1'^{(l)}}.$$

Then, using a controlled-not operation between Bob's bits (control) and the difference vector (target), we produce l qubits containing the same bit values as Alice's, with an arbitrarily large probability:

$$|e_1^{(l)}\rangle_{e_1^{(l)}} |s_1^{(l)} \oplus e_1^{(l)}\rangle_{s_1'^{(l)}} \to |e_1^{(l)}\rangle_{e_1^{(l)}} |s_1^{(l)}\rangle_{s_1'^{(l)}}.$$

This is how the l systems s_1' are created.

Following this approach for the next slices, we can define:

$$|\bar{s}\rangle_{\bar{s}} |s_1\rangle_{\mathsf{s}_1'} |E_1(y,\bar{s})\rangle_{e_1} |\bar{E}_1(y,\bar{s})\rangle_{\bar{e}_1}$$
$$\to |\bar{s}\rangle_{\bar{s}} |s_1\rangle_{\mathsf{s}_1'} |E_1(y,\bar{s})\rangle_{e_1} |E_2(y,\bar{s},s_1)\rangle_{e_2} |\bar{E}_2(y,\bar{s},s_1)\rangle_{\bar{e}_2},$$

and reasonably assume that the bit value given in s_1' is equal to Alice's $S_1(x)$. This reasoning can be applied iteratively, so as to fill the system $\mathsf{s}_{1...m}'$ with all the corrected bit values, and with an extra step to set $\bar{E}(y,\bar{s},s_{1...m})$ in $\bar{\mathsf{e}}$.

As a last step, Bob can revert the ECC decoding operations and come back to the situation where he has blank qubits in $\mathsf{s}_{1...m}'$ as depicted in Fig. 12.4.

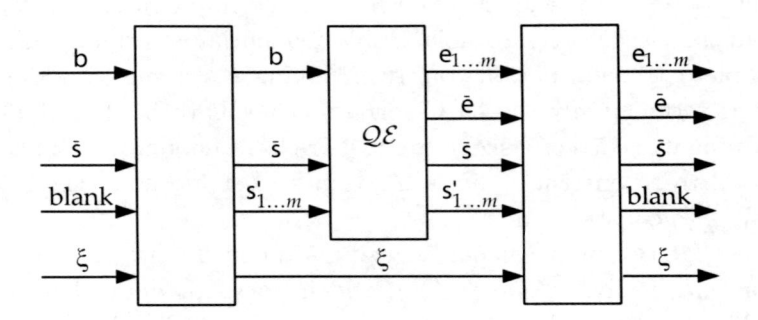

Fig. 12.4. Schematic description of $\mathcal{QE}$ and the use of the systems $\mathsf{s}_{1...m}'$. Reprinted with permission from [175] © 2005 by the American Physical Society.

Finally, the qubits produced by $\mathcal{QE}$ can be transformed into EPR pairs using the CSS codes and the syndromes Alice sent to Bob.

Phase coherence

Neither the unitary transformation $\mathcal{QS}$ nor $\mathcal{QE}$ take into account the modulation of the coherent state in the **p**-quadrature. By ignoring what happens in the a_2 system of Eq. (12.2), the reduced system $\rho_{\mathsf{a}_1\mathsf{b}}$ has an undetermined position on the p axis of Bob's side and thus lacks phase coherence:

$$\rho_{\mathsf{a_1 b}} = \int \mathrm{d}x\mathrm{d}x'\mathrm{d}p \sqrt{p_{N(0,\Sigma_A\sqrt{N_0})}(x) p_{N(0,\Sigma_A\sqrt{N_0})}(x') p_{N(0,\Sigma_A\sqrt{N_0})}(p)}$$
$$|x\rangle_{\mathsf{a_1}} \langle x'|D(\mathrm{i}p)|x+\mathrm{i}0\rangle_{\mathsf{b}}\langle x'+\mathrm{i}0|D^\dagger(\mathrm{i}p),$$

with $D(\mathrm{i}p) = \mathrm{e}^{\mathrm{i}p\mathsf{x}/4N_0}$ the displacement operator.

To ensure the phase coherence and thus reduce the phase error rate, we assume that Alice also sends the a_2 system to Bob, just like she does for the $\bar{\mathsf{s}}$ system and the syndromes, since the modulation in the p-quadrature is independent of the key. Bob can take it into account before applying $\mathcal{QE}$, by displacing his state along the p-quadrature in order to bring it on the x-axis.

Actually, we could formally include this a_2-dependent operation in the $\mathcal{QE}$ mapping, by adding $|p\rangle_{\mathsf{a}_2}$ to its input and output (unmodified) and by multiplying by a factor of the form $\mathrm{e}^{\mathrm{i}p\mathsf{y}/4N_0}$ in Eq. (12.5), with N_0 the vacuum fluctuations. For notation simplicity, however, I have mentioned it here without explicitly writing it.

Also, for the simplicity of the notation in the next section, we can assume without loss of generality that the coefficients of $|\Psi\rangle$ in the x-basis of b are real, after adjustment by Bob as a function of p.

Construction of $\bar{S}$ and $\bar{E}$

Let us now make the construction of the functions $\bar{S}$ and $\bar{E}$ explicit. First assume, for simplicity, that we have only one slice ($m=1$) – for this we do not write the slice index as a subscript. The mapping has thus the following form:

$$|x\rangle_{\mathsf{a_1}}|y\rangle_{\mathsf{b}}$$
$$\to \sigma(x)|S(x)\rangle_{\mathsf{s}}|\bar{S}(x)\rangle_{\bar{\mathsf{s}}}\epsilon(y,\bar{S}(x),S(x))|E(y,\bar{S}(x))\rangle_{\mathsf{e}}|\bar{E}(y,\bar{S}(x),S(x))\rangle_{\bar{\mathsf{e}}},$$

where $\sigma(x) = (\mathrm{d}_x \bar{S})^{-1/2}(x)$, $\epsilon(y,\bar{s},s) = (\partial_y \bar{E})^{-1/2}(y,\bar{s},s)$, and $\bar{S}$ and $\bar{E}$ range between 0 and 1.

Let us take some state ρ of the systems $\mathsf{s\bar{s}e\bar{e}}$. In the entanglement purification picture, our goal is to be able to extract entangled pairs in the subsystem $\rho_{\mathsf{se}} = \mathrm{Tr}_{\mathrm{All}\backslash\{\mathsf{s,e}\}}(\rho)$. We thus want ρ to be a product state of the form $\rho_{\mathsf{se}} \otimes \rho_{\bar{\mathsf{s}}\bar{\mathsf{e}}}$. If $\bar{S}(X)$ contains information about $S(X)$, or if $\bar{E}(Y,\bar{S}(X),S(X))$ contains information about $E(Y,\bar{S}(X))$, the subsystem ρ_{se} will not be pure. In the prepare-and-measure picture, information on $S(X)$ in $\bar{S}(X)$ will be known to Eve and therefore may not be considered as secure. Note that information in $\bar{E}(\dots)$ is not disclosed, but since it is excluded from the subsystems from which we wish to extract entanglement (or secrecy), any correlation with $\bar{\mathbf{e}}$ will reduce the number of entangled qubits (or secret bits); or stated otherwise, the calculated number of secret bits will be done as if $\bar{E}(\dots)$ were public. As an extreme example, if $S(X)$ and $E(Y, \bar{S}(X))$ are perfectly correlated and if $S(X)$ can be found directly as a function of $\bar{S}(X)$, then ρ_{se} will be of the form $\rho_{\mathsf{se}} = p_0|00\rangle\langle 00| + p_1|11\rangle\langle 11|$, which does not allow us to extract any EPR pairs, or equivalently, does not contain any secret information. Consequently, $\bar{S}$ and $\bar{E}$ should be as statistically independent as possible of S and E.

We define $\bar{S}$ and $\bar{E}$ as the following cumulative probability functions: $\bar{S}(x) = \Pr[X \leq x \mid S(X) = S(x)]$ and $\bar{E}(y, \bar{s}, s) = \Pr[Y \leq y \mid \bar{S}(X) = \bar{s}, S(X) = s, E(Y, \bar{s}) = E(y, \bar{s})]$. By definition, these functions are uniformly distributed between 0 and 1, each independently of the other variables available to the party calculating it (Alice for $\bar{S}$ and Bob for $\bar{E}$). These functions also enjoy the property of making the subsystem ρ_{se} pure in absence of eavesdropping (i.e., when ρ is pure), indicating that this choice of $\bar{S}$ and $\bar{E}$ does not introduce more impurity in ρ_{se} than ρ already has.

Theorem 14 ([175]) *Let ρ be a pure state. With the functions $\bar{S}$ and $\bar{E}$ defined as above, the subsystem ρ_{se} is also pure.*

When more than one slice is involved, the functions $\bar{S}$ and $\bar{E}$ are defined similarly:

$$\bar{S}(x) = \Pr[X \leq x | S_{1\dots m}(X) = S_{1\dots m}(x)], \tag{12.6}$$

$$\bar{E}(y, \bar{s}, s_{1\dots m}) = \Pr[Y \leq y | \bar{S}(X) = \bar{s} \wedge S_{1\dots m}(X) = s_{1\dots m}$$
$$\wedge E_1(Y, \bar{s}) = E_1(y, \bar{s}) \wedge \cdots \wedge E_m(Y, \bar{s}, s_{1\dots m-1}) = E_m(y, \bar{s}, s_{1\dots m-1})]. \tag{12.7}$$

Retro-compatibility of privacy amplification

As noted in Section 12.2.3 for the case of BB84, the privacy amplification derived from CSS codes still allows us to use a randomly selected hash function from any universal family of linear hash functions. Let us discuss what is possible in the case of the quantum sliced error correction.

With classical sliced error correction, we produce m bits per key element, which altogether enter the selected hash function. In contrast, the quantum generalization implies that we estimate different phase error rates e_i^{ph} for each slice $i \in \{1\dots m\}$. Since each slice uses an independent CSS codes, we thus expect to input each slice to an independent hash function. This is

12.3 Application to the GG02 protocol

indeed correct, but I now show that the two options are equivalent in this scope.

Theorem 15 *Processing each slice i using an independent linear universal family of hash functions of input size l and output size k_i can be implemented by processing ml bits in a hash function selected from $\mathcal{H}_{\mathrm{GF}(2^{ml}) \to \{0,1\}^{\sum k_i}}$ and selecting non-overlapping subsets of k_i output bits for each slice $i \in \{1 \ldots m\}$.*

Proof

For simplicity, let us assume that we use only $m = 2$ slices, namely $x_1, x_2 \in \mathrm{GF}(2)^l$. With independent privacy amplification, the privacy of slice $i \in \{1, 2\}$ is amplified with a hash function determined by some $k_i \times l$ matrix $H_i'^{\mathrm{ph}}$ coming from $\mathcal{H}_i = \mathcal{H}_{\mathrm{GF}(2^l) \to \{0,1\}^{k_i}}$, and the output bits are determined by $H_i'^{\mathrm{ph}} x_i$. It is clear that the family $\mathcal{H}_i$ is still universal if the output bits are translated by some vector δ_i independent of x_i, i.e., giving $H_i'^{\mathrm{ph}} x_i + \delta_i$.

Let us now consider the case of all slices being processed together. The input bits are concatenated to form the vector $x = (x_1 x_2)$ and the hash function is represented by some $(k_1 + k_2) \times 2l$ matrix H'^{ph}. This matrix can be expanded as

$$H'^{\mathrm{ph}} = \begin{pmatrix} H_{11}'^{\mathrm{ph}} & H_{12}'^{\mathrm{ph}} \\ H_{21}'^{\mathrm{ph}} & H_{22}'^{\mathrm{ph}} \end{pmatrix},$$

with $H_{ij}'^{\mathrm{ph}}$ of size $k_i \times l$. The first k_1 bits are $H_{11}'^{\mathrm{ph}} x_1 + H_{12}'^{\mathrm{ph}} x_2$, while the following k_2 bits are $H_{21}'^{\mathrm{ph}} x_1 + H_{22}'^{\mathrm{ph}} x_2$. As stated in the definition of $\mathcal{H}_{\mathrm{GF}(2^l) \to \{0,1\}^k}$ in Section 7.2, any set of bits can be extracted from the product in $\mathrm{GF}(2^{2l})$, hence both matrices $(H_{11}'^{\mathrm{ph}} H_{12}'^{\mathrm{ph}})$ and $(H_{21}'^{\mathrm{ph}} H_{22}'^{\mathrm{ph}})$ make universal families of hash functions. By identifying $H_{11}'^{\mathrm{ph}} = H_1'^{\mathrm{ph}}$ and $H_{12}'^{\mathrm{ph}} x_2 = \delta_1$, we find that the first k_1 bits are calculated as if x_1 is processed by an independent hash function. The same conclusion follows with the next k_2 bits and x_2. Of course, the same reasoning can be applied for any number m of slices. □

12.3.5 The attenuation channel

I now apply the slicing construction and display some results on the rates one can achieve in an important practical case. These results serve as an example and do not imply an upper bound on the achievable rates or distances. Instead, they can be viewed as lower bounds on an achievable secure rate in the particular case of an attenuation channel with given losses. Stated otherwise, this section simulates the rates we would obtain in a real experiment where Alice and Bob would be connected by an attenuation channel. For more general properties of the construction, please refer to Section 12.3.6.

The purpose of this section is twofold. First, I wish to illustrate the idea of the previous section and show that it serves realistic practical purposes. Beyond the generality of the sliced error correction, its implementation may be easier than it first appears. Furthermore, the purification (or distillation) of more than one qubit (or bit) per sample is useful, as illustrated below.

Second, it is important to show that the construction works in a case as important as the attenuation channel. Clearly, requesting that a QKD protocol yields a non-zero secret key rate under all circumstances is unrealistic – an eavesdropper can always block the entire communication. On the other hand, a QKD protocol that would always tell Alice and Bob that zero secure bits are available would be perfectly secure but obviously also completely useless. Of course, between these two extreme situations, the practical efficiency of a QKD protocol is thus important to consider.

The attenuation channel can be modeled as if Eve installed a beam splitter in between two sections of a lossless line, sending a vacuum state at the second input. We here assume that Alice sends coherent states with a modulation variance of $\Sigma_A^2 N_0 = 31 N_0$, which gives Alice and Bob up to $I(X;Y) = 2.5$ common bits in absence of losses or noise. This matches the order of magnitude implemented in [77]. We define the slices S_1 and S_2 by dividing the real axis into four equiprobable intervals labeled by two bits, with S_1 representing the least significant bit and S_2 the most significant one. More precisely, $S_1(x) = 0$ when $x \leq -\tau$ or $0 < x \leq \tau$ and $S_1(x) = 1$ otherwise, with $\tau = \sqrt{2 \times 31 N_0}\, \mathrm{erf}^{-1}(1/2)$, and $S_2(x) = 0$ when $x \leq 0$ and $S_2(x) = 1$ otherwise.

In this constructed example, we wish to calculate the theoretical secret key rate we would obtain in an identical setting. For various loss values, the secret key rates are evaluated by numerically calculating $\mathrm{Tr}((\mathbf{Z}_{s_i} \otimes \mathbf{Z}_{e_i})\rho)$, to obtain the bit error rates of slices $i = 1, 2$ and $\mathrm{Tr}((\mathbf{X}_{s_i} \otimes \mathbf{X}_{e_i})\rho)$ to obtain the phase error rates. Then, assuming asymptotically efficient binary codes, the rate is $R = R_1 + R_2 = \sum_{i=1,2}(1 - h(e_i^{\mathrm{bit}}) - h(e_i^{\mathrm{ph}}))$.

Using this two-slice construction, it is possible to obtain the EPR rates described in Table 12.1. For the case with no losses, it is thus possible to distill $R = 0.752 + 0.938 = 1.69$ EPR pairs per sample. Also, note that the phase error rate increases faster with the attenuation for ρ_2 than for ρ_1, with $\rho_i = \rho_{s_i e_i} = \mathrm{Tr}_{\mathrm{All}\setminus\{s_i, e_i\}}(\rho)$. This intuitively follows from the fact that the

12.3 Application to the GG02 protocol

information Eve can gain from the output of her beam splitter first affects the most significant bit contained in $S_2(X)$.

Table 12.1. *Error and EPR rates with two slices in an attenuation channel.*

Losses	e_1^{bit} (ρ_1)	e_1^{ph}	R_1	e_2^{bit} (ρ_2)	e_2^{ph}	R_2
0.0 dB	3.11%	0.53%	0.752	0.0000401	0.710%	0.938
0.4 dB	3.77%	13.7%	0.193	0.0000782	28.6%	0.135
0.7 dB	4.32%	20.0%	0.0204	0.000125	37.5%	0.0434
1.0 dB	–	–	–	0.000194	42.3%	0.0147
1.4 dB	–	–	–	0.000335	45.6%	0.00114

Reprinted with permission from [175] © 2005 by the American Physical Society.

Because of the higher bit error rate in ρ_1, it is not possible to distill EPR pairs in slice 1 with losses beyond 0.7 dB with this construction. It is, however, still possible to distill EPR pairs in slice 2, up to 1.4 dB losses (about 10 km with fiber optics with losses of 0.15 dB/km). This result does not pose any fundamental limit, as it can vary with the modulation variance and with the choice of the functions S_1 and S_2. Note that the slice functions could be optimized in various ways, one of which being to use other intervals (as in Section 9.3.2, not necessarily equiprobable and possibly chosen as a function of the losses), and another being to consider multi-dimensional.

Note that although this example involves a Gaussian channel, this property is not exploited here and such a calculation can be as easily done for a non-Gaussian attack.

When only individual attacks are taken into account (see, for instance, Section 9.3.3), the reconciliation efficiency allows one to obtain higher secret key rates. The intervals in the classical version of the SEC have been optimized and are shaped to maximize the efficiency, whereas the intervals used in this section are chosen to be equiprobable. This fixed choice explains a part of the difference in efficiencies. The reason for the intervals to be equiprobable is not arbitrary; the correlations between slices have to be minimized.

In the EPP picture, each slice is purified independently. From the point of view of the purification of a slice i, any correlation outside the system $s_i e_i$ leads to phase errors and a decrease in the secret key rate. This argument also applies between the slices. Each slice is like an independent instance of BB84. Even if the other slices are not part of the eavesdropper's knowledge, the computation of the secret key rate for one slice is such that the correlations with another slice is indistinguishable from eavesdropping; if slices i and j are correlated, slice i will have phase errors due to slice j and vice versa. But it works both ways and the penalty is, in a sense, counted twice: both the rate of slice i and of slice j are decreased because of the correlations. The equiprobable intervals are chosen to avoid such costly correlations.

12.3.6 Asymptotic behavior

In this section, I describe the behavior of the slice construction when the slice and slice estimator mappings take as input a block of d states, with d arbitrarily large. In Section 9.1.2, the classical sliced error correction is shown to reduce to Slepian–Wolf coding [167] (asymmetric case with side information) when using asymptotically large block sizes. Here we study the quantum case, which is different at least by the fact that privacy amplification is explicitly taken into account.

For simplicity of the notation, we will study the asymptotic behavior in the case of an individually-probed channel only – although Eve's measurement can be collective. A study of finite-width probing with a width w much smaller than the key size ($w \ll l$) would give the same results, since in both cases it allows us to consider a sequence of identical random experiments and to study the typical case. However, joint attacks, with width as large as the key size, are outside the scope of this section, as the statistical tools presented here would not be suitable.

It is important to stress that we investigate here what the secret key rates would be if the actual channel were an individually-probed one. The use of the protocol of this section still requires the evaluation of the phase error rate in all cases, and this quantity is sufficient to determine the number of secret key bits. In the case of more general joint attacks, the secret key rates stated in the special cases below would then differ from the one obtained using the phase error rate.

Direct reconciliation

We consider a block of d states and the functions S, $\bar{S}$, E and $\bar{E}$ on blocks of d variables as well. Then we make d grow to infinity.

We define the following state, with the action of the channel modeled as joining system **b** with that of an eavesdropper Eve, and with p left out as a public classical parameter:

$$|\Psi(p)\rangle = \int \mathrm{d}x\, g(x)|x\rangle_{a_1}|\phi(x,p)\rangle_{b,\text{eve}}. \tag{12.8}$$

We consider d such states coherently, and the mappings $\mathcal{QS}$ and $\mathcal{QE}$ take all d states as input. We will follow the lines of the reasoning in [53, 83, 157] to show that the secret key rate tends to $I(X;Y) - I'_{AE}$ for $d \to \infty$, with X the random variable representing Alice's measure of $\mathbf{a}_1$ with $\mathbf{x}$, Y the measure of $\mathbf{b}$ with $\mathbf{x}$, and $I'_{AE} = H(X) + H(\rho_{\text{eve}}) - H(\rho_{a_1,\text{eve}})$, where $H(\rho)$ is the von Neumann entropy of a state ρ.

When $d \to \infty$, the qubits produced by $\mathcal{QS}$ can be split into three categories. First, there are a certain number of qubits, the disclosed value of which allows Alice and Bob to correct (almost) all bit errors for the remaining slices. Then, among the remaining slices, a certain number of qubits allow Alice and Bob to correct (almost) all phase errors for the rest of the qubits. These last qubits are thus equivalent to secret key bits in the prepare-and-measure protocol. This is the idea behind the following theorem.

Theorem 16 ([175]) *For d sufficiently large, there exist slice and slice estimator operators $\mathcal{QS}$ and $\mathcal{QE}$, operating on groups of d states, such that the secret key rate can be as close as desired to $I(X;Y) - I'_{AE}$.*

Note that in the particular case of the attenuation channel, an evaluation of the secret key rate can be found in [80, 133].

Reverse reconciliation

So far, we have always assumed that the slices apply to Alice and the slice estimators to Bob. However, we also have to consider the case of reverse reconciliation, as it resists losses better than direct reconciliation – see Section 11.3.2.

Let us start again from the state $|\Psi(p)\rangle$ as in Eq. (12.8), and rewrite the state $|\phi(x,p)\rangle_{b,\text{eve}}$ as

$$|\phi(x,p)\rangle_{b,\text{eve}} = \int \mathrm{d}y\, f(x,p,y)|y\rangle_b |\phi(x,p,y)\rangle_{\text{eve}}.$$

Let $h(y,p)$ be a non-negative real function such that

$$h^2(y,p) = \int \mathrm{d}x |g(x,p)f(x,p,y)|^2.$$

Then,
$$|\Psi(p)\rangle = \int dy\, h(y,p)|y\rangle_{\sf b}|\phi'(y,p)\rangle_{{\sf a}_1,{\sf eve}},$$
with
$$|\phi'(y,p)\rangle_{{\sf a}_1,{\sf eve}} = \int dx\, g(x,p) f(x,p,y)/h(y,p)|x\rangle_{{\sf a}_1}|\phi(x,p,y)\rangle_{\sf eve}.$$

Thus, by applying the same argument as for direct reconciliation, we can asymptotically reach $I(X;Y) - I'_{\rm BE}$ secret bits when $\mathcal{QS}$ is applied on system ${\sf b}$ and $\mathcal{QE}$ on system ${\sf a}_1$, with $I'_{\rm BE} = H(Y) + H(\rho_{\sf eve}) - H(\rho_{{\sf b},{\sf eve}})$. The evaluation of the secret key rate for reverse reconciliation can also be found in [80, 133], which indicates that such a quantity is always strictly positive in the case of an attenuation channel, regardless of the losses, for a sufficiently large modulation variance.

In [80], Grosshans evaluates the Holevo information for collective eavesdropping both in direct and reverse reconciliation cases. Gaussian channels are investigated, including the attenuation channel. An important point is that, for reverse reconciliation, non-zero secret key rates can still be obtained whatever the transmission losses, for a sufficiently large modulation variance. This result has direct consequences on the result of this section, as the EPP achieves the same rate in the same circumstances. Asymptotically, basing the secret key rate on the phase error rate does not impose a penalty when the channel is individually-probed.

12.3.7 Discussion

Before concluding this section, let me discuss some aspects of the above protocol.

Retro-compatibility

What are the differences between the above protocol and the original GG02 protocol?

Regarding Alice's modulation, nothing changes: she can still use a Gaussian modulation of coherent states. On Bob's side, the tomography requires the ability to measure any quadrature $\mathbf{x}_\theta$ for $0 \leq \theta < 2\pi$, not only $\mathbf{x}$ and $\mathbf{p}$ as in Section 11.3. This seems technically reasonable and only requires an extra phase modulator at Bob's side.

The secret-key distillation may seem different, as we use the generalization of sliced error correction, but the techniques used are, in fact, identical. As noted in Section 12.2.3, any linear correction protocol can be used for reconciliation. For privacy amplification, we saw that the family of hash functions must be linear. In particular, $\mathcal{H}_{\text{GF}(2^l) \to \{0,1\}^k}$, described in Section 7.3, can be used for privacy amplification, and it was proved in Theorem 15 that all bits produced by sliced error correction can be processed together, as has already been done in Section 11.4. Of course, the number of bits to be removed is estimated via the phase error rate and not using the mutual information.

To sum up, the differences between this EPP-derived protocol and the original protocol are the following:

- The channel must be probed using tomography. For this, Bob must be able to perform homodyne detection in *all quadratures*.
- The privacy amplification must use a *linear* universal family of hash functions.
- The number of bits to sacrifice is determined by the *phase error rate*, as estimated by tomography.

Advantages and disadvantages

Besides the fact that this generalization of GG02 is robust against joint eavesdropping attacks, let us discuss the other advantages and disadvantages of the proposed method.

A strong advantage of using the equivalence with EPP is the adaptability of the obtained results. The analysis above is not specific to the fact that Alice's modulation is Gaussian. We could easily replace the state $|\Psi\rangle$ defined in Eq. (12.2) by a state that represents another modulation. The resulting secret key rate would have to be recalculated, but the analysis would still be applicable. For instance, Alice's modulation is not perfectly Gaussian in practice. Then again, the amplitude modulator cannot extend infinitely, as would be required by a Gaussian distribution, but the distribution is truncated. Also, the generated modulation values do not vary continuously but are generated by a digital–analog converter. Although these effects seem negligible, there is no proof that they are actually negligible on the secrecy of the final key, especially in complex eavesdropping conditions. QKD derived from EPP provides us with a fairly general tool that enables us to integrate such effects into the state $|\Psi\rangle$ and thus in the evaluation of the phase error rate.

An advantage of the method is its asymptotic efficiency: the protocol based on EPP is as efficient as the original GG02 protocol when the channel is individually-probed.

Unfortunately, the secret key rates of Section 12.3.5 turn out to be lower than those of Section 9.3.3, in both cases using one-dimensional slicing. Using the phase error rates impose a penalty on the results, which can however be improved using multidimensional slicing.

Another disadvantage of the method is that it requires tomography of the quantum channel. The tomography may be unrealistic in practice. The generality of the reconciliation in our case implies that, to ensure that the phase error rate is estimated with a small variance, it is expected that Alice and Bob would need to use a large number n of samples. In principle, one could use an arbitrarily large block size l such that $l \gg n$ and the estimation part becomes negligible.

12.4 Conclusion

In this chapter, I first overviewed the important eavesdropping strategies and related them to secret key distillation. For collective and joint eavesdropping strategies, the secret-key distillation techniques must be somehow adapted. A way to do this is to study the equivalence of QKD and entanglement distillation protocols.

I then detailed the equivalence between EPP and the BB84 protocol and translate secret-key distillation into entanglement purification with CSS codes.

Finally, I studied the equivalence between EPP and the GG02 protocol. I generalized sliced error correction to make the coherent-state protocol resistant to joint eavesdropping strategies.

Appendix
Symbols and abbreviations

Alice	the legitimate party who sends quantum states to Bob
Bob	the legitimate party who receives quantum states from Alice
Claude	another name for either Alice or Bob. In Chapter 5, Claude sends commands to Dominique. In Chapters 8 and 9, the final key depends on Claude's key elements; for direct reconciliation, Claude=Alice; for reverse reconciliation, Claude=Bob.
Dominique	another name for either Alice or Bob. In Chapter 5, Dominique responds to the commands sent by Claude. In Chapters 8 and 9, Dominique aligns his/her key elements on Claude's; for direct reconciliation, Dominique=Bob; for reverse reconciliation, Dominique=Alice.
Eve	the eavesdropper and active enemy of Alice and Bob
AES	advanced encryption standard
BB84	the protocol of Bennett and Brassard published in 1984 [10]
BBBSS	the protocol of Bennett, Bessette, Brassard, Salvail and Smolin [13]
BCP	binary correction protocol
BEC	binary erasure channel
BSC	binary symmetric channel
BSKDP	block secret-key distillation protocol
CA	certification authority
CMAC	command message authentication code
CP	completely positive
CRMAC	command-response message authentication code
CSKDP	continuous secret-key distillation protocol
CSS	Calderbank–Shor–Steane

DES	data encryption standard
ECC	error correcting code
ECM	entangling cloning machine
EPP	entanglement purification protocol
EPR	Einstein–Podolsky–Rosen
IEC	interactive error correction
iff	if and only if
FFT	fast Fourier transform
FY	the protocol of Furukawa and Yamazaki [62]
GG02	the protocol of Grosshans and Grangier published in 2002 [74]
GP01	the protocol of Gottesman and Preskill published in 2001 [68]
LDPC	low-density parity check
LLR	log-likelihood ratio
LO	local oscillator
MAC	message authentication code
MAP	maximum a posteriori
MSD	multistage soft decoding
NLUI	near-lossless unrestricted inputs
NTT	number-theoretic transform
PK	public key
PKI	public key infrastructure
PNS	photon-number splitting
QKD	quantum key distribution
RI	restricted inputs
SEC	sliced error correction
SKD	secret-key distillation
snr	signal-to-noise ratio
UI	unrestricted inputs
$\oplus$	the mod 2 bitwise addition of bit strings or bit vectors
$|\cdot|$	the length of a string or the size of a set
$|n\rangle$	a photon number state with n photons
$|\alpha\rangle$	a coherent state centered on $(\operatorname{Re}\alpha, \operatorname{Im}\alpha)$
$|\alpha, s\rangle$	a squeezed state centered on $(\operatorname{Re}\alpha, \operatorname{Im}\alpha)$ with squeezing s
$\langle \cdot \rangle$	the expected value of the observable in argument
$\mathbf{C}$	the set of complex numbers
e	the error rate
$E[\cdot]$	the expected value of the argument
G	the transmission gain in a quantum channel

$\mathrm{GF}(q)$	the finite (Galois) field of size q, with q a prime or a prime power
$h(p)$	the entropy of a binary random variable with distribution $\{p, 1-p\}$
$H(X)$	the Shannon entropy of X, if X is a discrete random variable, or the differential entropy of X if X is continuous
$H(\rho)$	the von Neumann entropy of ρ
$H_r(X)$	the order-r Rényi entropy of X
$I(X;Y)$	the mutual information between X and Y
$\mathbf{I}$	the identity operator
l	the block size; the number of key elements processed through secret-key distillation
M	all the messages exchanged during reconciliation, with $|M|$ their total size
n	the number of pulses used for the estimation of the quantum channel
$N(\mu,\sigma)$	a Gaussian distribution with mean μ and standard deviation σ
$\mathbf{N}$	the set of natural numbers
N_0	the variance of vacuum fluctuations
$\mathbf{p}, \mathbf{x}$	the quadrature amplitude operators
$\Pr[\cdot]$	the probability of an event
r	the number of rows of a parity-check matrix or (often equivalently) the number of bits disclosed by a binary reconciliation protocol
$\mathbf{R}$	the set of real numbers
T	the transmission probability in a quantum channel
Tr	the trace of an operator
X	Claude's key elements
$\mathbf{X}, \mathbf{Y}, \mathbf{Z}$	the Pauli operators or Pauli matrices
Y	Dominique's key elements
Z	Eve's knowledge on the key elements
$\mathbf{Z}$	the set of integral numbers
$\mathbf{Z}_n$	the set $\{0, 1, \ldots, n-1\}$ with $\mathrm{mod}\, n$ addition and multiplication
$\mathbf{Z}_p^*$	the set $\{1, 2, \ldots, p-1\}$ with $\mathrm{mod}\, p$ multiplication
ξ	the syndrome of an error-correcting code
χ	the noise variance in the quantum channel in N_0 units
Ψ	the key elements after reconciliation

Bibliography

[1] A. Aaron and B. Girod, Compression with side information using turbo codes, *Proc. IEEE Data Compression Conf. (DCC)* (2002).

[2] R. Alléaume, F. Treussart, G. Messin *et al.*, Experimental open-air quantum key distribution with a single-photon source, *New J. Phys.* **6**, 92 (2004).

[3] N. Alon and A. Orlitsky, Source coding and graph entropies, *IEEE Trans. Inform. Theory* **42**, 1329–1339 (1996).

[4] J. Arndt, *Algorithms for programmers (working title)*, draft version of 2004-July-13, http://www.jjj.de/fxt/ (2004).

[5] L. R. Bahl, J. Cocke, F. Jelinek and J. Raviv, Optimal decoding of linear codes for minimizing symbol error rate, *IEEE Trans. Inform. Theory* **IT-20**, 284–287 (1974).

[6] H. J. Beker, Analog speech security systems, *Advances in Cryptology – Eurocrypt '82*, Lecture Notes in Computer Science, 130–146 (1982).

[7] M. Bellare, R. Canetti and H. Krawczyk, Keying hash functions for message authentication, *Advances in Cryptology – Crypto '96*, Lecture Notes in Computer Science, 1–15 (1996).

[8] M. Ben-Or, M. Horodecki, D. W. Leung, D. Mayers and J. Oppenheim, The universal composable security of quantum key distribution, *Proc. Second Theory of Cryptography Conf. (TCC)*, 386–406 (2005).

[9] K. Bencheikh, T. Symul, A. Jankovic and J.-A. Levenson, Quantum key distribution with continuous variables, *J. Mod. Opt.* **48**, 1903–1920 (2001).

[10] C. H. Bennett and G. Brassard, Public-key distribution and coin tossing, *Proceedings of the IEEE International Conference on Computers, Systems, and Signal Processing, Bangalore, India*, 175–179 (1984).

[11] C. H. Bennett, G. Brassard and J. Robert, Privacy amplification by public discussion, *SIAM J. Comput.* **17**, 210–229 (1988).

[12] C. H. Bennett, G. Brassard and A. K. Ekert, Quantum cryptography, *Sci. Am.* **267**, 50–57 (1992).

[13] C. H. Bennett, F. Bessette, G. Brassard, L. Salvail and J. Smolin, Experimental quantum cryptography, *J. Cryptol.* **5**, 3–28 (1992).

[14] C. H. Bennett, Quantum cryptography using any two non-orthogonal states, *Phys. Rev. Lett.* **68**, 3121–3124 (1992).

[15] C. H. Bennett, G. Brassard, C. Crépeau and U. M. Maurer, Generalized privacy amplification, *IEEE Trans. Inform. Theory* **41**, 1915–1923 (1995).

[16] C. H. Bennett, G. Brassard, S. Popescu *et al.*, Purification of noisy entan-

glement and faithful teleportation via noisy channels, *Phys. Rev. Lett.* **76**, 722–725 (1996).

[17] C. Berrou, A. Glavieux and P. Thitimajshima, Near Shannon limit error-correction coding and decoding: turbo-codes, *Proc. Int. Conf. Communications*, 1064–1070 (1993).

[18] D. S. Bethune, M. Navarro and W. P. Risk, Enhanced autocompensating quantum cryptography system, *Appl. Opt. LP* **41**, 1640–1648 (2002).

[19] A. Beveratos, R. Brouri, T. Gacoin et al., Single photon quantum cryptography, *Phys. Rev. Lett.* **89**, 187901 (2002).

[20] E. Biham and A. Shamir, Differential cryptanalysis of DES-like cryptosystems, *Advances in Cryptology – Crypto '90*, Lecture Notes in Computer Science, 2–21 (1990).

[21] E. Biham and A. Shamir, Differential cryptanalysis of DES-like cryptosystems, *J. Cryptol.* **4**, 3–72 (1991).

[22] E. Biham and A. Shamir, Differential cryptanalysis of the full 16-round DES, *Advances in Cryptology – Crypto '92*, Lecture Notes in Computer Science, 494–502 (1992).

[23] E. Biham, M. Boyer, P. O. Boykin, T. Mor and V. Roychowdhury, A proof of the security of quantum key distribution, *Proc. of the 32nd Annual ACM Symp. on Theory of Computing*, 715–724 (1999).

[24] J. Black, S. Halevi, H. Krawczyk, T. Krovetz and P. Rogaway, UMAC: fast and secure message authentication, *Advances in Cryptology – Crypto '99*, Lecture Notes in Computer Science, 216–233 (1999).

[25] M. Bloch, A. Thangaraj and S. W. McLaughlin, Efficient reconciliation of correlated continuous random variables using LDPC codes, *arXiv e-print* cs.IT/0509041 (2005).

[26] M. Bourennane, F. Gibson, A. Karlsson et al., Experiments on long-wavelength (1550 nm) 'plug and play' quantum cryptography systems, *Opt. Express* **4**, 383–387 (1999).

[27] G. Brassard and L. Salvail, Secret-key reconciliation by public discussion, *Advances in Cryptology – Eurocrypt '93*, Lecture Notes in Computer Science, 410–423 (1993).

[28] B. P. Brent, S. Larvala and P. Zimmermann, A fast algorithm for testing reducibility of trinomials mod 2 and some new primitive trinomials of degree 3021377, *Math. Comp.* **72**, 1443–1452 (2003).

[29] B. P. Brent, S. Larvala and P. Zimmermann, *Search for primitive trinomials (mod 2)*, http://web.comlab.ox.ac.uk/oucl/work/richard.brent/trinom.html (2004).

[30] D. Bruß, M. Cinchetti, G. M. D'Ariano and C. Macchiavello, Phase covariant quantum cloning, *Phys. Rev. A* **62**, 012302 (2000).

[31] W. T. Buttler, S. K. Lamoreaux, J. R. Torgerson et al., Fast, efficient error reconciliation for quantum cryptography, *Phys. Rev. A* **67**, 052303 (2003).

[32] C. Cachin, *Entropy Measures and Unconditional Security in Cryptography*, Ph.D. thesis, ETH Zürich (1997).

[33] A. R. Calderbank, Multilevel codes and multistage decoding, *IEEE Trans. Inform. Theory* **37**, 222–229 (1989).

[34] A. R. Calderbank and P. W. Shor, Good quantum error-correcting codes exist, *Phys. Rev. A* **54**, 1098–1105 (1996).

[35] J. Cardinal, *Quantization With an Information-Theoretic Distortion Mea-*

sure, Technical Report 491, Université Libre de Bruxelles (2002).
[36] J. Cardinal and G. Van Assche, Construction of a shared secret key using continuous variables, *Proc. IEEE Inform. Th. Workshop (ITW)* (2003).
[37] J. Cardinal, S. Fiorini and G. Van Assche, On minimum entropy graph colorings, *Proc. IEEE International Symposium on Information Theory (ISIT)* (2004).
[38] J. L. Carter and M. N. Wegman, Universal classes of hash functions, *J. Comput. Syst. Sci.* **18**, 143–154 (1979).
[39] N. J. Cerf, A. Ipe and X. Rottenberg, Cloning of continuous quantum variables, *Phys. Rev. Lett.* **85**, 1754–1757 (2000).
[40] N. J. Cerf, M. Lévy and G. Van Assche, Quantum distribution of Gaussian keys using squeezed states, *Phys. Rev. A* **63**, 052311 (2001).
[41] G. Chiribella, G. M. D'Ariano, P. Perinotti and N. J. Cerf, Extremal quantum cloning machines, *Phys. Rev. A* **72**, 042336 (2005).
[42] M. Christandl, R. Renner and A. Ekert, A generic security proof for quantum key distribution, *arXiv e-print* `quant-ph/0402131` (2004).
[43] ComScire – Quantum World Corp., *Design Principles and Testing of the QNG Model J1000KU^{TM}*, http://www.comscire.com/Products/J1000KU/ (2005).
[44] T. H. Cormen, C. E. Leiserson, R. L. Rivest and C. Stein, *Introduction to Algorithms*, second edition, Cambridge, MA, MIT Press (2001).
[45] N. T. Courtois and J. Pieprzyk, Cryptanalysis of block ciphers with overdefined systems of equations, *Advances in Cryptology – Asiacrypt 2002*, Lecture Notes in Computer Science, 267–287 (2002).
[46] T. M. Cover and J. A. Thomas, *Elements of Information Theory*, New York, Wiley & Sons (1991).
[47] P. Crescenzi, V. Kann, M. Halldórsson, M. Karpinski and G. Woeginger, *A Compendium of NP Optimization Problems*, http://www.nada.kth.se/~viggo/problemlist/compendium.html (2005).
[48] I. Csiszár and J. Körner, Broadcast channels with confidential messages, *IEEE Trans. Inform. Theory* **24**, 339–348 (1978).
[49] K. M. Cuomo and A. V. Oppenheim, Circuit implementation of synchronized chaos with applications to communications, *Phys. Rev. Lett.* **71**, 65–68 (1993).
[50] G. M. D'Ariano, C. Macchiavello and N. Sterpi, Systematic and statistical errors in homodyne measurements of the density matrix, *Quantum Semiclass. Opt.* **9**, 929–939 (1997).
[51] G. M. D'Ariano, M. G. Paris and M. F. Sacchi, Quantum tomography, *Adv. Imag. Elect. Phys.* **128**, 205–308 (2003).
[52] J. Daemen and V. Rijmen, *The Design of Rijndael*, Berlin, Springer-Verlag (2002).
[53] I. Devetak and A. Winter, Relating quantum privacy and quantum coherence: an operational approach, *Phys. Rev. Lett.* **93**, 080501 (2004).
[54] W. Diffie and M. E. Hellman, New directions in cryptography, *IEEE Trans. Inform. Theory* **22**, 644–654 (1976).
[55] B. Efron and R. J. Tibshirani, *An Introduction to the Bootstrap*, London, CRC Press LLC (1998).
[56] A. Einstein, B. Podolsky and N. Rosen, Can quantum-mechanical description of physical reality be considered complete?, *Phys. Rev.* **47**, 777–780 (1935).
[57] C. Elliott, Building the quantum network, *New J. Phys.* **4**, 46 (2002).
[58] C. Elliott, D. Pearson and G. Troxel, Quantum cryptography in practice,

2003 Conference on Applications, Technologies, Architectures, and Protocols for Computer Communications, 227–238 (2003).

[59] C. Ellison, C. Hall, R. Milbert and B. Schneier, Protecting secret keys with personal entropy, *Future Gener. Comp. Syst.* **16**, 311–318 (2000).

[60] Federal Information Processing Standard (FIPS), *Data Encryption Standard*, Publication 46, Washington D.C., National Bureau of Standards, U.S. Department of Commerce (1977).

[61] C. A. Fuchs, N. Gisin, R. B. Griffiths, C.-S. Niu and A. Peres, Optimal eavesdropping in quantum cryptography. I Information bound and optimal strategy, *Phys. Rev. A* **56**, 1163–1172 (1997).

[62] E. Furukawa and K. Yamazaki, Application of existing perfect code to secret key reconciliation, *Conf. Proc. Int. Symp. Commun. Inform. Tech.*, 397–400 (2001).

[63] R. G. Gallager, Low density parity check codes, *IRE Trans. Inf. Theory* **IT-8**, 21–28 (1962).

[64] N. Gisin, G. Ribordy, W. Tittel and H. Zbinden, Quantum cryptography, *Rev. Mod. Phys.* **74**, 145–195 (2002).

[65] N. Gisin, S. Fasel, B. Kraus, H. Zbinden and G. Ribordy, Trojan horse attacks on quantum key distribution systems, *arXiv e-print* `quant-ph/0507063` (2005).

[66] C. Gobby, Z. L. Yuan and A. J. Shields, Quantum key distribution over 122 km of standard telecom fiber, *Appl. Phys. Lett.* **84**, 3762–3864 (2004).

[67] S. Goldwasser and S. Micali, Probabilistic encryption, *J. Comp. Syst. Sci.* **28**, 270–299 (1984).

[68] D. Gottesman and J. Preskill, Secure quantum key distribution using squeezed states, *Phys. Rev. A* **63**, 022309 (2001).

[69] D. Gottesman and H.-K. Lo, Proof of security of quantum key distribution with two-way classical communications, *IEEE Trans. Inform. Theory* **49**, 457–475 (2003).

[70] D. Gottesman, H.-K. Lo, N. Lütkenhaus and J. Preskill, Security of quantum key distribution with imperfect devices, *Quantum Inf. Comput.* **4**, 325–360 (2004).

[71] F. Gray, *Pulse Code Communication*, United States Patent 2,632,058 (March 17, 1953).

[72] R. M. Gray and D. L. Neuhoff, Quantization, *IEEE Trans. Inform. Theory* **44**, 2325–2383 (1998).

[73] W. Greiner, *Quantum Mechanics—An Introduction*, third edition, Berlin, Springer-Verlag (1994).

[74] F. Grosshans and P. Grangier, Continuous variable quantum cryptography using coherent states, *Phys. Rev. Lett.* **88**, 057902 (2002).

[75] F. Grosshans and P. Grangier, Reverse reconciliation protocols for quantum cryptography with continuous variables, *arXiv e-print* `quant-ph/0204127` (2002).

[76] F. Grosshans, *Communication et Cryptographie Quantiques avec des Variables Continues*, Ph.D. thesis, Université Paris XI (2002).

[77] F. Grosshans, G. Van Assche, J. Wenger *et al.*, Quantum key distribution using Gaussian-modulated coherent states, *Nature* **421**, 238–241 (2003).

[78] F. Grosshans, J. Wenger, R. Brouri, N. J. Cerf and P. Grangier, Virtual entanglement and reconciliation protocols for quantum cryptography with continuous variable, *Quantum Inf. Comput.* **3**, 535–552 (2003).

[79] F. Grosshans and N. J. Cerf, Continuous-variable quantum cryptography is secure against non-Gaussian attacks, *Phys. Rev. Lett* **92**, 047905 (2004).
[80] F. Grosshans, Collective attacks and unconditional security in continuous variable quantum key distribution, *Phys. Rev. Lett.* **94**, 020504 (2005).
[81] C. G. Günther, An identity-based key-exchange protocol, *Advances in Cryptology – Eurocrypt '89*, Lecture Notes in Computer Science, 29–37 (1989).
[82] M. Hillery, Quantum cryptography with squeezed states, *Phys. Rev. A* **61**, 022309 (2000).
[83] A. S. Holevo, The capacity of the quantum channel with general signal states, *IEEE Trans. Inform. Theory* **44**, 269–273 (1998).
[84] Horst Görtz Institute for IT-Security and Institute for Applied Information Processing and Communications, *Fourth Conference on the Advanced Encryption Standard (AES)* (2004).
[85] D. A. Huffman, A method for the construction of minimum redundancy codes, *Proc. IRE*, 1098–1101 (1952).
[86] R. J. Hughes, W. T. Buttler, P. G. Kwiat *et al.*, Practical quantum cryptography for secure free-space communications, *arXiv e-print* **quant-ph/9905009** (1999).
[87] R. J. Hughes, J. E. Nordholt, D. Derkacs and C. G. Peterson, Practical free-space quantum key distribution over 10 km in daylight and at night, *New J. Phys* **4**, 43 (2002).
[88] W.-Y. Hwang, Quantum key distribution with high loss: toward global secure communication, *Phys. Rev. Lett.* **91**, 057901 (2003).
[89] S. Iblisdir, G. Van Assche and N. J. Cerf, Security of quantum key distribution with coherent states and homodyne detection, *Phys. Rev. Lett.* **93**, 170502 (2004).
[90] id Quantique, $Quantis^{TM}$, http://www.idquantique.com/ (2005).
[91] id Quantique, $Vectis^{TM}$, http://www.idquantique.com/ (2005).
[92] D. Kahn, *The Codebreakers*, Scribner (1996).
[93] S. C. Kak, Scrambling and randomization, *Advances in Cryptology – Crypto '81*, Lecture Notes in Computer Science, 59–63 (1981).
[94] J. Kelsey, B. Schneier, D. Wagner and C. Hall, Side channel cryptanalysis of product ciphers, *J. Comput. Secur.* **8**, 141–158 (2000).
[95] A. Kerckhoffs, La cryptographie militaire, *Journal des sciences militaires* **IX**, 5–38 (1883).
[96] D. E. Knuth, *The Art of Computer Programming*, vol. 2, third edition, Reading, MA, Addison-Wesley (1997).
[97] P. Kocher, J. Jaffe and B. Jun, Differential power analysis, *Advances in Cryptology – Crypto '99*, Lecture Notes in Computer Science, 388–397 (1999).
[98] R. König, U. Maurer and R. Renner, Privacy amplification secure against an adversary with selectable knowledge, *Proc. IEEE International Symposium on Information Theory (ISIT)* (2004).
[99] R. König, U. Maurer and R. Renner, On the power of quantum memory, *IEEE Trans. Inform. Theory* **51**, 2391–2401 (2005).
[100] J. Körner and A. Orlitsky, Zero-error information theory, *IEEE Trans. Inform. Theory* **44**, 2207–2229 (1998).
[101] P. Koulgi, E. Tuncel, S. Regunathan and K. Rose, Minimum redundancy zero-error source coding with side information, *Proc. IEEE International Symposium on Information Theory (ISIT)* (2001).
[102] P. Koulgi, E. Tuncel and K. Rose, On zero-error coding of correlated sources,

Proc. IEEE International Symposium on Information Theory (ISIT), 62 (2002).
[103] P. Koulgi, E. Tuncel, S. Regunathan and K. Rose, On zero-error source coding with decoder side information, *IEEE Trans. Inform. Theory* **49**, 99–111 (2003).
[104] H. Krawczyk, LFSR-based hashing and authentication, *Advances in Cryptology – Crypto '94*, Lecture Notes in Computer Science, 129–139 (1994).
[105] C. Kurtsiefer, P. Zarda, M. Halder et al., A step towards global key distribution, *Nature* **419**, 450 (2002).
[106] G. G. Langdon, An introduction to arithmetic coding, *IBM J. Res. Dev.* **28**, 135–149 (1984).
[107] A. D. Liveris, Z. Xiong and C. N. Georghiades, Compression of binary sources with side information at the decoder using LDPC codes, *IEEE Commun. Lett.* **6**, 440–442 (2002).
[108] H.-K. Lo and H. F. Chau, Unconditional security of quantum key distribution over arbitrarily long distances, *Science* **283**, 2050–2056 (1999).
[109] H.-K. Lo, Method for decoupling error correction from privacy amplification, *New J. Phys.* **5**, 36 (2003).
[110] H.-K. Lo, X. Ma and K. Chen, Decoy state quantum key distribution, *Phys. Rev. Lett.* **94**, 230504 (2005).
[111] J. Lodewyck, T. Debuisschert, R. Tualle-Brouri and P. Grangier, Controlling excess noise in fiber-optics continuous-variable quantum key distribution, *Phys. Rev. A* **72**, 050303(R) (2005).
[112] S. Lorenz, N. Korolkova and G. Leuchs, Continuous variable quantum key distribution using polarization encoding and post selection, *Appl. Phys. B* **79**, 273–277 (2004).
[113] MagiQ Technologies, QPN^{TM}, http://www.magiqtech.com/ (2005).
[114] Y. Mansour, N. Nisan and P. Tiwari, The computational complexity of universal hash functions, *Theor. Comput. Sci.* **107**, 121–133 (1993).
[115] M. Matsui, Linear cryptanalysis method for DES cipher, *Advances in Cryptology – Eurocrypt '93*, Lecture Notes in Computer Science, 386–397 (1993).
[116] M. Matsui, The first experimental cryptanalysis of the data encryption standard, *Advances in Cryptology – Crypto '94*, Lecture Notes in Computer Science, 1–11 (1994).
[117] U. M. Maurer, Secret key agreement by public discussion from common information, *IEEE Trans. Inform. Theory* **39**, 733–742 (1993).
[118] U. Maurer, Information-theoretically secure secret-key agreement by NOT authenticated public discussion, *Advances in Cryptology – Eurocrypt '97*, Lecture Notes in Computer Science, 209–225 (1997).
[119] U. M. Maurer and S. Wolf, The intrinsic conditional mutual information and perfect secrecy, *Proc. IEEE International Symposium on Information Theory (ISIT)*, 88 (1997).
[120] U. M. Maurer and S. Wolf, Unconditionally secure key agreement and the intrinsic conditional information, *IEEE Trans. Inform. Theory* **45**, 499–514 (1999).
[121] U. M. Maurer, Authentication theory and hypothesis testing, *IEEE Trans. Inform. Theory* **46**, 1350–1356 (2000).
[122] U. Maurer and S. Wolf, Information-theoretic key agreement: from weak to strong secrecy for free, *Advances in Cryptology – Eurocrypt 2000*, Lecture Notes in Computer Science, 351–368 (2000).

[123] U. Maurer and S. Wolf, Secret-key agreement over unauthenticated public channels—part I: definitions and a completeness Result, *IEEE Trans. Inform. Theory* **49**, 822–831 (2003).

[124] U. Maurer and S. Wolf, Secret-key agreement over unauthenticated public channels—part II: the simulatability condition, *IEEE Trans. Inform. Theory* **49**, 832–838 (2003).

[125] U. Maurer and S. Wolf, Secret-key agreement over unauthenticated public channels—part III: privacy amplification, *IEEE Trans. Inform. Theory* **49**, 839–851 (2003).

[126] D. Mayers, Unconditional security in quantum cryptography, *J. ACM* **48**, 351–406 (2001).

[127] A. J. Menezes, P. C. van Oorschot and S. A. Vanstone, *Handbook of Applied Cryptography*, London, CRC Press LLC (1997).

[128] A. Muller, H. Zbinden and N. Gisin, Quantum cryptography over 23 km in installed under-lake telecom fibre, *Europhys. Lett.* **33**, 335 (1996).

[129] J. Muramatsu, Secret key agreement from correlated source outputs using LDPC matrices, *Proc. IEEE International Symposium on Information Theory (ISIT)* (2004).

[130] J. Muramatsu, T. Uyematsu and T. Wadayama, Low-density parity-check matrices for coding of correlated sources, *IEEE Trans. Inform. Theory* **51**, 3645–3654 (2005).

[131] S. Murphy and M. J. Robshaw, Essential algebraic structure within the AES, *Advances in Cryptology – Crypto 2002*, Lecture Notes in Computer Science, 1–16 (2002).

[132] National Institute of Standards and Technology (NIST), *Advanced Encryption Standard*, `http://www.nist.gov/aes` (2000).

[133] M. Navascués and A. Acín, Security bounds for continuous variables quantum key distribution, *Phys. Rev. Lett.* **94**, 020505 (2005).

[134] P. Navez, Statistical confidentiality tests for a quantum transmission using continuous variables, *Eur. Phys. J. D* **18**, 219–228 (2002).

[135] K.-C. Nguyen, *Extension des Protocoles de Réconciliation en Cryptographie Quantique*, Masters thesis, Université Libre de Bruxelles (2002).

[136] K.-C. Nguyen, G. Van Assche and N. J. Cerf, Side-information coding with turbo codes and its application to quantum key distribution, *Proc. International Symposium on Information Theory and its Applications (ISITA)* (2004).

[137] K.-C. Nguyen, G. Van Assche and N. J. Cerf, One-dimensional and multi-dimensional reconciliation using turbo codes for quantum key distribution, *Proc. 26th Symp. on Inform. Theory in the Benelux* (2005).

[138] M. A. Nielsen and I. L. Chuang, *Quantum Computation and Quantum Information*, Cambridge, Cambridge University Press (2000).

[139] N. Nisan and D. Zuckerman, Randomness is linear in space, *J. Comput. Syst. Sci.* **52**, 43–52 (1996).

[140] K. G. Paterson, F. Piper and R. Schack, Why quantum cryptography?, *arXiv e-print* `quant-ph/0406147` (2004).

[141] Y. Peres, Iterating von Neumann's procedure for extracting random bits, *Annals of Statistics* **20**, 590–597 (1992).

[142] J. Preskill, *Lecture Notes for Physics 229: Quantum Information and Computation*, `http://theory.caltech.edu/~preskill/ph229/` (1998).

[143] Protego, *R200-USB^{TM} TRNG Product Specification*,

http://www.protego.se/sg200_d.htm (2005).

[144] T. C. Ralph, Continuous variable quantum cryptography, *Phys. Rev. A* **61**, 010303 (2000).

[145] R. Raz, O. Reingold and S. Vadhan, Extracting all the randomness and reducing the error in Trevisan's extractors, *Proc. Symp. Theory of Computing*, 149–158 (1999).

[146] M. D. Reid, Quantum cryptography with a predetermined key, using continuous-variable Einstein–Podolsky–Rosen correlations, *Phys. Rev. A* **62**, 062308 (2000).

[147] R. Renner and S. Wolf, New bounds in secret-key agreement: the gap between formation and secrecy extraction, *Advances in Cryptology – Eurocrypt 2003*, Lecture Notes in Computer Science, 562–577 (2003).

[148] A. Rényi, On measures of entropy and information, *Proc. 4th Berkeley Symp. on Math. Statistics and Prob.*, 547–561 (1961).

[149] G. Ribordy, J. Gautier, H. Zbinden and N. Gisin, Fast and user-friendly quantum key distribution, *J. Mod. Opt.* **47**, 517–531 (2000).

[150] R. L. Rivest, A. Shamir and L. M. Adleman, A method for obtaining digital signatures and public-key cryptosystems, *Commun. ACM* **21**, 120–126 (1978).

[151] B. Sanders, J. Vučković and P. Grangier, Single photons on demand, *Europhys. News* **36**, 56–58 (2005).

[152] D. C. Schmidt, *The ADAPTIVE Communication Environment (ACE^{TM})*, http://www.cs.wustl.edu/~schmidt/ACE.html (2005).

[153] B. Schneier, *Applied Cryptography*, second edition, New York, John Wiley & Sons (1995).

[154] A. Schönhage and V. Strassen, Schnelle Multiplikation großer Zahlen, *Computing* **7**, 281–292 (1971).

[155] A. Schönhage, Schnelle Multiplikation von Polynomen über Körpern der Charakteristik 2, *Acta Inform.* **7**, 395–398 (1977).

[156] B. Schumacher, Quantum coding, *Phys. Rev. A* **51**, 2738–2747 (1995).

[157] B. Schumacher and M. D. Westmoreland, Sending classical information via noisy quantum channels, *Phys. Rev. A* **56**, 131–138 (1997).

[158] M. O. Scully and M. S. Zubairy, *Quantum Optics*, Cambridge, Cambridge University Press (1997).

[159] C. E. Shannon, *Analogue of the Vernam System for Continuous Time Series*, Memorandum MM 43-110-44, Bell Laboratories (1943).

[160] C. E. Shannon, A mathematical theory of communication, *Bell Syst. Tech. J.* **27**, 623–656 (1948).

[161] C. E. Shannon, Communication theory of secrecy systems, *Bell Syst. Tech. J.* **28**, 656–715 (1949).

[162] P. W. Shor, Algorithms for quantum computation: discrete logarithms and factoring, *Proceedings of the 35th Symposium on Foundations of Computer Science*, 124–134 (1994).

[163] P. W. Shor and J. Preskill, Simple proof of security of the BB84 quantum key distribution protocol, *Phys. Rev. Lett.* **85**, 441–444 (2000).

[164] C. Silberhorn, N. Korolkova and G. Leuchs, Quantum key distribution with bright entangled beams, *Phys. Rev. Lett.* **88**, 167902 (2002).

[165] C. Silberhorn, T. C. Ralph, N. Lütkenhaus and G. Leuchs, Continuous variable quantum cryptography – beating the 3 dB loss limit, *Phys. Rev. Lett.* **89**, 167901 (2002).

[166] S. Singh, *The Code Book*, New York, Doubleday (1999).
[167] D. Slepian and J. K. Wolf, Noiseless coding of correlated information sources, *IEEE Trans. Inform. Theory* **19**, 471–480 (1973).
[168] A. Steane, Multiple particle interference and quantum error correction, *Proc. R. Soc. London A* **452**, 2551–2577 (1996).
[169] D. R. Stinson, Universal hashing and authentication codes, *Design. Code. Cryptogr.* **4**, 369–380 (1994).
[170] D. R. Stinson, *Cryptography Theory and Practice*, second edition, London, CRC Press LLC (2002).
[171] D. Stucki, N. Gisin, O. Guinnard, G. Ribordy and H. Zbinden, Quantum key distribution over 67 km with a plug & play system, *New J. Phys* **4**, 41 (2002).
[172] L. Trevisan, Extractors and pseudorandom generators, *J. ACM* **48**, 860–879 (2001).
[173] G. Van Assche, J. Cardinal and N. J. Cerf, Reconciliation of a quantum-distributed Gaussian key, *IEEE Trans. Inform. Theory* **50**, 394–400 (2004).
[174] G. Van Assche, *Information-Theoretic Aspects of Quantum Key Distribution*, Ph.D. thesis, Université Libre de Bruxelles (2005).
[175] G. Van Assche, S. Iblisdir and N. J. Cerf, Secure coherent-state quantum key distribution protocols with efficient reconciliation, *Phys. Rev. A* **71**, 052304 (2005).
[176] G. S. Vernam, Cipher printing telegraph systems for secret wire and radio telegraphic communications, *J. IEEE* **55**, 109–115 (1926).
[177] U. Wachsmann, R. F. Fischer and J. B. Huber, Multilevel codes: theoretical concepts and practical design rules, *IEEE Trans. Inform. Theory* **45**, 1361–1391 (1999).
[178] E. Waks, K. Inoue, C. Santori *et al.*, Quantum cryptography with a photon turnstile, *Nature* **420**, 762 (2002).
[179] J. Walker, *HotBits: Genuine Random Numbers, Generated by Radioactive Decay*, http://www.fourmilab.ch/hotbits/ (2005).
[180] D. F. Walls and G. J. Milburn, *Quantum Optics*, Berlin, Springer-Verlag (1994).
[181] X.-B. Wang, Beating the photon-number-splitting attack in practical quantum cryptography, *Phys. Rev. Lett.* **94**, 230503 (2005).
[182] M. N. Wegman and J. L. Carter, New hash functions and their use in authentication and set equality, *J. Comput. Syst. Sci.* **22**, 265–279 (1981).
[183] E. W. Weisstein, *Number Theoretic Transform*, from MathWorld – a Wolfram web resource, http://mathworld.wolfram.com/NumberTheoreticTransform.html (2005).
[184] E. W. Weisstein, *Gray Code*, from MathWorld – a Wolfram web resource, http://mathworld.wolfram.com/GrayCode.html (2005).
[185] J. Wenger, *Dispositifs Impulsionnels pour la Communication Quantique à Variables Continues*, Ph.D. thesis, Université Paris XI (2004).
[186] S. Wiesner, Conjugate coding, *Sigact News* **15**, 78–88 (1983).
[187] H. S. Witsenhausen, The zero-error side information problem and chromatic numbers, *IEEE Trans. Inform. Theory* **22**, 592–593 (1976).
[188] T. Wörz and J. Hagenauer, Iterative decoding for multilevel codes using reliability information, *Proc. IEEE Globecom Conference*, 1779–1784 (1992).
[189] A. D. Wyner, The wire-tap channel, *Bell Syst. Tech. J.* **54**, 1355–1387 (1975).
[190] Y.-O. Yan and T. Berger, On instantaneous codes for zero-error coding of

two correlated sources, *Proc. IEEE International Symposium on Information Theory (ISIT)* (2000).

[191] H. Zbinden, J. Gautier, N. Gisin *et al.*, Interferometry with Faraday mirrors for quantum cryptography, *Electron. Lett.* **33**, 586–588 (1997).

[192] H. Zbinden, H. Bechmann-Pasquinucci, N. Gisin and G. Ribordy, Quantum cryptography, *Appl. Phys. B* **67**, 743–748 (1998).

[193] Q. Zhao and M. Effros, Optimal code design for lossless and near lossless source coding in multiple access networks, *Proc. IEEE Data Compression Conf. (DCC)*, 263–272 (2001).

[194] Q. Zhao, S. Jaggi and M. Effros, Side information source coding: low-complexity design and source independence, *Proc. Asilomar Conference on Signals and Systems* (2002).

[195] Q. Zhao and M. Effros, Low complexity code design for lossless and near-lossless side information source codes, *Proc. IEEE Data Compression Conf. (DCC)* (2003).

[196] Q. Zhao and M. Effros, Lossless and near-lossless source coding for multiple access networks, *IEEE Trans. Inform. Theory* **49**, 112–128 (2003).

Index

a-priori information, 134, 149
advanced encryption standard, 20, 23, 25
afterpulse, 170
arithmetic coding, 120, 122
authentication, 26, 28, 96

BB84, 159
beam splitter, 163, 165, 167, 191, 192, 202, 238
Bell basis, 53, 211
binary interactive error correction, 124, 200
birefringence, 163, 169
bit error rate, 11, 212
block cipher, 20, 27, 76

Cascade, 125, 128, 200, 201, 218
certification authority, 32, 73
channel
    binary erasure channel, 41
    binary symmetric channel, 41, 144, 200
    Gaussian channel, 46, 186, 190
channel capacity, 41
channel coding, 41
chosen plaintext, 21
chromatic number, 119
ciphertext, 16
cloning
    asymmetric Gaussian cloning machines, 188
    cloning machine, 58, 172, 173, 187
    entangling cloning machine, 192, 201, 203
    no-cloning theorem, 58
    no-cloning uncertainty relation, 188, 192
    phase-covariant cloning machine, 173
code
    instantaneous, 37
    non-singular, 37
    prefix-free, 37, 118, 121
    uniquely decodable, 37
coherent state, 59, 160, 184, 189, 196, 225
    weak coherent state, 160, 217, 219
coloring, 119
    entropy of a coloring, 119
    minimum cardinality coloring, 119, 120
    minimum entropy coloring, 120

commutator, 51
composability, 206, 220
computational basis, 52
computational security, 25, 28, 30
confusable, 117
convolutional code, 129

dark count, 162, 170, 176, 178
data encryption standard, 20, 23–25
decoy state method, 180, 219
density matrix, 54
differential cryptanalysis, 21
Diffie–Hellman key agreement, 31
direct reconciliation, 191, 192, 198, 231
distributed source coding, 114

eavesdropping
    collective eavesdropping, 206, 221, 242
    individual eavesdropping, 170, 187, 191, 199, 202, 205
    individual Gaussian eavesdropping, 192
    joint eavesdropping, 206, 207, 221, 222
    non-Gaussian finite-width eavesdropping, 221
efficiency, 170
entangled state, 55, 192
    entangled photon pair, 161, 209
    maximally entangled states, 53, 208, 223
entanglement purification, 207, 225
entropy, 36
    chromatic entropy, 119, 120
    conditional entropy, 40
    differential entropy, 45
    entropy of a coloring, 119
    Rényi entropy, 44, 89
    von Neumann entropy, 55
error-correcting code, 42
    convolutional code, 129
    linear error-correcting code, 42
    low-density parity-check code, 137
    orthogonal code, 43
    turbo code, 129, 135
estimation, 11, 97, 212

exhaustive search, 19, 23, 25
extractor, 91
extrinsic information, 136, 149

fidelity, 55, 56, 58, 173, 208, 221
Fock basis, 59, 165, 179, 229

Gaussian
   Gaussian channel, 46
   Gaussian distribution, 186, 190
   Gaussian modulation, 184, 223, 242
   Gaussian random variable, 46, 60, 225
   Gaussian state, 59
GG02, 189, 221
graph
   confusability graph, 119, 121, 123
   probabilistic graph, 119
   Tanner graph, 138

Heisenberg uncertainty principle, 51, 58, 184, 190
homodyne detection, 59, 194, 223
Huffman coding, 36, 119, 122

impersonation, 27, 28
intercept and resend, 171

Kerckhoffs' principle, 16
key, 4, 16
   private key, 29
   public key, 29
   secret key, 17, 18, 20, 27, 28
key element, 7, 85
   binary key element, 159
   Gaussian key element, 184, 189, 222
keyspace, 20, 25, 28
keystream, 18
known plaintext, 22, 24

linear cryptanalysis, 24
linear feedback shift register, 18, 130
local oscillator, 194
log-likelihood ratio, 133, 137
low-density parity-check code, 137

Mach–Zehnder interferometer, 165
   double Mach–Zehnder interferometer, 166, 167
man-in-the-middle, 26, 32
Markov chain, 43
measurement, 7, 54
message authentication code, 27, 28
message digest function, 30
mixed state, 54
multi-photon pulse, 161, 177, 178, 219
mutual information, 41
   intrinsic mutual information, 95

near-lossless unrestricted inputs code, 118
number-theoretic transform, 109

observable, 51

one-time pad, 16, 201
open air, 162, 164
operator, 51
   annihilation operator, 59
   bit syndrome operator, 211
   bit-flip operator, 53, 226, 232
   creation operator, 59
   displacement operator, 235
   Pauli operators, 52, 226
   phase syndrome operator, 211
   phase-flip operator, 53, 226, 232
   photon number operator, 59, 187
   quorum of operators, 227

parametric downconversion, 161
parity bits, 130
peer review, 25
perfect forward secrecy, 74
perfect secrecy, 16, 17, 25, 47, 98
phase, 165
phase error rate, 212
photon detector, 163, 165, 167, 169
photon number states, 59, 179
plaintext, 16
plug-and-play, 167
polarization, 58, 162, 167
prefix, 37
   proper prefix, 37
prepare-and-measure, 208
privacy amplification, 11, 85, 215
public classical authenticated channel, 5, 85, 113, 230, 232
public-key cryptography, 29, 31
public-key infrastructure, 32, 73
pure state, 54

quadrature amplitude, 58, 184, 225, 229
quantum channel, 167, 185
   attenuation channel, 191, 192, 194
   quantum Gaussian channel, 191, 239
quantum key distribution, 4
quantum tomography, 57, 227
qubit, 6, 52, 53

random variable
   binary random variable, 114
   continuous random variable, 12, 45, 98, 114
   discrete random variable, 12, 87, 114
rate, 36, 40, 41
   secret key rate, 94
reconciliation, 11, 85
   efficiency of reconciliation, 116
   interactive reconciliation, 114, 115, 199, 218
   one-way reconciliation, 114, 115
restricted inputs code, 118
reverse reconciliation, 191, 192, 198, 202, 241
Rijndael, 21
round function, 21
RSA, 29–31

Schumacher compression, 55

Index

secrecy capacity, 93
secret-key cryptography, 15, 26, 31
secret-key distillation, 11, 85, 198, 207
seperable state, 55
sifting, 8, 159, 175, 186
signature, 30
single-photon source, 162, 178, 219
slice, 142, 230, 232
   slice estimator, 142, 230, 233
source coding, 35, 37
source coding with side information, 114, 116
   near-lossless, 117
   zero-error, 117, 119
squeezed state, 60, 184, 185, 223
stream cipher, 18
strong forward secrecy, 74
strongly universal family of hash functions, 67
substitution, 27, 28
superposition, 7, 50
syndrome, 42, 127, 211
systematic bits, 130

tagged bits, 180, 219
trace, 54
   partial trace, 54
triple-DES, 20
turbo code, 129, 135, 203

unitary operator, 52
universal family of hash functions, 28, 88, 101, 199, 215
unrestricted inputs code, 118

vacuum fluctuations, 58
vacuum state, 59, 191, 238
variational distance, 91

Wigner function, 60
Winnow, 127, 128

Lightning Source UK Ltd.
Milton Keynes UK
UKOW020127230313

208047UK00003B/34/P

9 781107 410633